Springer Theses

Recognizing Outstanding Ph.D. Research

Aims and Scope

The series "Springer Theses" brings together a selection of the very best Ph.D. theses from around the world and across the physical sciences. Nominated and endorsed by two recognized specialists, each published volume has been selected for its scientific excellence and the high impact of its contents for the pertinent field of research. For greater accessibility to non-specialists, the published versions include an extended introduction, as well as a foreword by the student's supervisor explaining the special relevance of the work for the field. As a whole, the series will provide a valuable resource both for newcomers to the research fields described, and for other scientists seeking detailed background information on special questions. Finally, it provides an accredited documentation of the valuable contributions made by today's younger generation of scientists.

Theses are accepted into the series by invited nomination only and must fulfill all of the following criteria

- They must be written in good English.
- The topic should fall within the confines of Chemistry, Physics, Earth Sciences, Engineering and related interdisciplinary fields such as Materials, Nanoscience, Chemical Engineering, Complex Systems and Biophysics.
- The work reported in the thesis must represent a significant scientific advance.
- If the thesis includes previously published material, permission to reproduce this must be gained from the respective copyright holder.
- They must have been examined and passed during the 12 months prior to nomination.
- Each thesis should include a foreword by the supervisor outlining the significance of its content.
- The theses should have a clearly defined structure including an introduction accessible to scientists not expert in that particular field.

More information about this series at http://www.springer.com/series/8790

Yang Liu

Multifunctional Gold Nanostars for Cancer Theranostics

Doctoral Thesis accepted by
the Duke University, USA

Springer

Author
Dr. Yang Liu
Department of Chemistry
Duke University
Durham, NC
USA

Supervisor
Prof. Tuan Vo-Dinh
Fitzpatrick Institute for Photonics,
Department of Biomedical Engineering,
Department of Chemistry
Duke University
Durham, NC
USA

ISSN 2190-5053 ISSN 2190-5061 (electronic)
Springer Theses
ISBN 978-3-030-09104-0 ISBN 978-3-319-74920-4 (eBook)
https://doi.org/10.1007/978-3-319-74920-4

Printed on acid-free paper

This Springer imprint is published by Springer Nature
The registered company is Springer International Publishing AG
The registered company address is: Gewerbestrasse 11, 6330 Cham, Switzerland

Supervisor's Foreword

Cancer is the most common cause of death in people aged less than 85 years, and it is estimated that in 2012, there were more than 14 million new cancer cases worldwide and 8 million deaths. Early detection and specific therapy has the potential to dramatically improve cancer patient's overall survival. Therefore, there is an urgent need to develop novel imaging and therapy methods to improve cancer management. Nanotheranostics, a combination of diagnostic and therapeutic modalities into a single nanoplatform, has great potential to be used for cancer management by offering early diagnostics, sensitive detection, real-time tracking, image-guided therapy, and therapeutic response monitoring. In particular, the use of nanoparticles in medicine has attracted increased attention for their unique efficacy and specificity in diagnostics and therapy. Among nanoparticles used for imaging and photothermal therapy, gold nanostars (GNS) are of particular interest as their structures have multiple sharp branches that produce the "lightning rod" effect that strongly enhances the local electromagnetic field and concentrates heating into cancer cells. The thesis work of Yang Liu involves the development and applications of a multifunctional GNS nanoprobe for sensitive cancer imaging and specific treatment, leading to important breakthroughs in biomedical studies. The multifunctional GNS nanoprobes can be used for (1) surface-enhanced Raman spectroscopy (SERS), (2) two-photon photoluminescence (TPL), (3) X-ray computed tomography (CT), (4) magnetic resonance imaging (MRI), (5) positron emission tomography (PET), and (6) photothermal therapy (PTT) for cancer treatment. The research demonstrated the GNS nanoprobe's capability for intracranial tumor detection down to submillimeter level in a murine model. This detection sensitivity is superior to any currently available noninvasive imaging modality. Current detectable tumor size limit using MRI is approximately 3–5 mm, which is not sufficiently sensitive for early brain tumor detection. Furthermore, the thesis research demonstrated that the use of plasmonic nanoparticles in a novel combination treatment referred to as Synergistic Immuno Photothermal Nanotherapy (SYMPHONY), which combines local photothermal therapy (PTT) using plasmonic GNS and immune checkpoint inhibitor-based immunomodulation, can enhance and broaden the effect of immune checkpoint inhibitors for use in treating both primary tumors as well as distal tumors and also produce an anticancer "vaccine" effect in a murine model. The important results on cancer photoimmunotherapy of this thesis have now been successfully extended into the treatment of a wide range of cancer diseases including breast cancer and brain cancer in murine models. This strategy could lead to an entirely new treatment paradigm that challenges traditional surgical resection approaches for many cancers and metastases.

Durham, NC, USA
January 2018

Tuan Vo-Dinh

Abstract

The prevalence of cancer has increasingly become a significant threat to human health and as such, there exists a strong need for developing novel methods for early detection and effective therapy. Nanotheranostics, a combination of diagnostic and therapeutic functions into a single nanoplatform, has great potential to be used for cancer management by allowing detection, real-time tracking, image-guided therapy, and therapeutic response monitoring. Gold nanostars (GNS) with tip-enhanced plasmonics have become one of the most promising platforms for cancer nanotheranostics. This work is aimed at addressing the challenges of sensitive cancer detection, metastasis treatment, and recurrence prevention by combining state-of-the-art nanotechnology, molecular imaging, and immunotherapy. A multifunctional GNS nanoprobe is developed with capabilities ranging from noninvasive, multi-modality cancer detection using positron emission tomography (PET), magnetic resonance imaging (MRI), and X-ray computed tomography (CT), to intraoperative tumor margin delineation with surface-enhanced Raman spectroscopy (SERS) and high-resolution nanoprobe tracking with two-photon photoluminescence (TPL), as well as cancer treatment with photoimmunotherapy. The GNS nanoprobe with PET scans is particularly exceptional in detecting brain malignancies as small as 0.5 mm. To the best of our knowledge, the developed GNS nanoprobe for PET imaging provides the most sensitive means of brain tumor detection reported so far. In addition, the GNS nanoprobe exhibits superior performance as photon-to-heat transducer and can be used for specific photothermal therapy (PTT). More importantly, GNS-mediated PTT combined with checkpoint inhibitor immunotherapy has been found to trigger a memorized immunoresponse to treat cancer metastasis and prevent recurrence in mouse model studies. Furthermore, a 6-month in vivo toxicity study including body weight monitoring, blood chemistry test, and histopathology examination demonstrates GNS nanoparticles' biocompatibility. Therefore, the multifunctional GNS nanoprobe exhibits superior cancer detection and treatment capabilities and has great promise for future clinical translation in cancer management.

Parts of this thesis have been published in the following journal articles:

[1] **Y. Liu**, P. Maccarini, G. M. Palmer, W. Etienne, Y. Zhao, C. T. Lee, X. Ma, B. A. Inman, T. Vo-Dinh. "Synergistic immuno photothermal nanotherapy (SYMPHONY) for the treatment of unresectable and metastatic cancer", Scientific Reports, 2017, 7, 8606: 1–6.

[2] **Y. Liu**, J. R. Ashton, E. J. Moding, H. Yuan, J. K. Register, J. Choi, M. J. Whitley, X. Zhao, Y. Qi, Y. Ma, G. Vaidyanathan, M. R. Zalutsky, D. G. Kirsch, C. T. Badea, T. Vo-Dinh, "A plasmonic gold nanostar theranostic probe for in vivo tumor imaging and photothermal therapy", Theranostics, 2015, 5(9), 946–960.

[3] **Y. Liu**, H. Yuan, F. R. Kersey, J. K. Register, M. C. Parrott, T. Vo-Dinh, "Plasmonic gold nanostars for multi-modality sensing and diagnostics", Sensors, 2015, 15(2), 3706–3720.

[4] **Y. Liu**, Z. Chang, H. Yuan, A. M. Fales and T. Vo-Dinh, "Quintuple-modality (SERS-MRI-CT-TPL-PTT) plasmonic nanoprobe for theranostics", Nanoscale, 2013, 5(21), 10127–10140.

[5] **Y. Liu**, H. Yuan, A. M. Fales and T. Vo-Dinh, "pH-sensing nanostar probe using surface-enhanced Raman scattering (SERS): theoretical and experimental studies", Journal of Raman Spectroscopy, 2013, 44(7), 980–986.

Acknowledgements

I really appreciate the opportunity of studying as a Ph.D. student at the Duke University and I feel proud of making my own contribution aimed to address challenges in cancer management. Only with all the help from my supervisor (Dr. Tuan Vo-Dinh), committee members (Dr. Benjamin Wiley, Dr. Jie Liu, and Dr. David Beratan), collaborators (Dr. David Kirsch, Dr. Michael Zalutsky, Dr. Cristian Badea, Dr. Ganesan Vaidyanathan, Dr. Hai Yan, Dr. Bennett Chin, Dr. Brant Inman, Dr. Gregory Palmer, Dr. Paolo Maccarini, and Dr. Gerald Grant), colleagues (Hsiangkuo Yuan, Andrew Fales, Hsin-Neng Wang, Janna Register, Stephen Norton, Guy Griffin, Bridget Crawford, Hoan Thanh Ngo, Ren Odion, Michelle Seywald, Austin Carpenter, Christopher Pirozzi, Zhengyuan Zhou, Paula Greer, Thomas Hawk, Wiguins Etienne, Yulin Zhao, Hengtao Zhang, Kelly Lee, Jeffrey Ashton, Everett Moding, Jaeyeon Choi, and Melodi Whitely), resources from Duke (Light Microscopy Core Facility, Optical Molecular Imaging and Analysis, Translational PET/CT Molecular Imaging Center and Shared Materials Instrumentation Facility), and many others, I can reach my fruitful achievements which I couldn't believe I would achieve them when I started my study at Duke in August 2010. I would also like to thank my family members for their tremendous physical and spiritual supports.

Thanks, everyone!

Contents

List of Figures

Chapter 1
Introduction

1.1 Specific Aims

Cancer has become a significant threat to human health with more than seven million deaths each year [1]. Early detection and specific therapy can significantly improve cancer patients' survival time but they are still unmet clinical needs calling for enormous efforts from both preclinical and clinical scientists [2]. This study is aimed at combining state-of-the-art nanotechnology, molecular imaging and immunotherapy for nanotheranostics, a combination of cancer therapeutics and diagnostics using a single multifunctional nanoprobe. The ideal nanotheranostics probe should be biocompatible and capable of targeting cancer, tracking biodistribution sensitively, and treating disease specifically [3, 4]. Multifunctional nanoprobes are superior to single function agents with the following advantages:

(1) enable image-guided treatment to improve therapeutic efficiency and specificity.
(2) can perform studies of nanoparticles' tumor uptake, pharmacokinetics and biodistribution without extra labeling.
(3) improve cost-effectiveness since several applications can be realized with one nanoprobe product.

In this study, it is hypothesized that the biocompatible multifunctional gold nanostars (GNS) probe can be used for both sensitive cancer imaging from the subcellular to whole body level as well as specific cancer treatment using photoimmunotherapy.

The following specific aims are proposed to investigate the above hypothesis:

Specific Aim 1: Development and optimization: Develop multimodality GNS nanoprobe for cancer theranostics: (a) Diagnostics: positron emission tomography (PET), magnetic resonance imaging (MRI), X-ray computed tomography (CT), two-photon photoluminescence (TPL) and surface-enhanced Raman spectroscopy (SERS); and (b) Therapeutics: photoimmunotherapy, a combination of photothermal

Y. Liu, *Multifunctional Gold Nanostars for Cancer Theranostics*, Springer Theses,
https://doi.org/10.1007/978-3-319-74920-4_1

therapy and immunotherapy. Perform in vivo evaluation to optimize the developed GNS theranostic nanoprobe.

Specific Aim 2: Diagnostics: Apply the multimodality GNS for in vivo cancer imaging and demonstrate the feasibility of submillimeter brain tumor detection with non-invasive imaging methods.

Specific Aim 3: Therapeutics: Apply the multimodality GNS for in vivo cancer therapy and demonstrate the feasibility of photoimmunotherapy to treat not only primary tumor but also cancer metastasis.

Successful completion of this study will generate a novel multifunctional GNS nanoprobe for applications in not only sensitive cancer detection at an early stage but also specific treatment for both primary tumor and cancer metastasis, all of which can provide great value for cancer management in future translational medicine applications.

1.2 Nanotheranostics

Nanotheranostics, where diagnostic and therapeutic functions are combined into a single nanometer-sized agent, provides an attractive means to improve cancer management by allowing tumor identification, real-time tracking, image-guided specific therapy, and continuous monitoring of therapeutic response [5, 6]. Gold nanoparticles (e.g., nanorods, nanoshells, nanocages, and nanostars) have received great attention as nanotheranostic agents due to their properties including strong localized surface plasmon resonance, in vivo stability, biocompatibility, and facile surface modification via strong thiol-gold chemical bonding [7–18]. They have been designed to have strong surface plasmon resonance in the near-infrared (NIR) region, which is considered to be the tissue optical window for biomedical imaging and therapy due to the low optical attenuation in this spectral range (700–900 nm) [19]. Those gold nanoparticles have been used for surface-enhanced Raman spectroscopy (SERS), multiphoton microscopy, photothermal therapy (PTT), photodynamic therapy (PDT), drug delivery and other applications [10, 20–23]. PTT uses gold nanoparticles as a transducer to convert NIR light to heat and generate a localized temperature increase [24, 25]. The distinct advantage of PTT with gold nanoparticles is that the NIR light only causes heating within tissues having gold nanoparticles accumulation, enabling specific treatment with minimal damage to surrounding normal tissues [13, 17, 26]. Tumors tend to have high uptake of nanoparticles due to the enhanced permeability and retention (EPR) effect [27–32]. As a result, they experience large temperature increases with NIR light exposure; surrounding tissues, which have minimal nanoparticle uptake, experience little change in temperature even with NIR irradiation. In addition, gold nanoparticles can also be used as contrast agents for X-ray computed tomography (CT) because gold has high X-ray attenuation ($\sim$2.7-times higher attenuation than iodine per mass) [33]. Furthermore, gold nanoparticles can be functionalized with a wide variety of molecules suitable for magnetic resonance imaging (MRI), positron emission tomography (PET) and optical imaging [4, 34–37]. Therefore, gold theranostic probes could be used to perform pre-treatment scans with PET, CT or MRI; intraoperative tumor detection with optical imaging and SERS; effective cancer PTT therapy with NIR light; targeted radiotherapy with loaded therapeutic radioisotopes; and therapeutic response monitoring with various imaging modalities.

Among those gold nanoparticles with different shapes, the GNS has unique tip enhanced plasmonics, which enable it to have superior electromagnetic field enhancement near their sharp spikes [38]. For example, it has been reported that GNS has extremely high TPL cross-section (more than 4.0×10^6 GM), which is 100 times higher than gold nanorod and 8000 times higher than gold nanocube [22]. The GNS has also been found to have higher photon-to-heat conversion efficiency than gold nanoshell (94% compared to 61%), which is currently under clinical trial for cancer photothermal therapy [24, 39]. Our group has developed a toxic surfactant-free method to synthesize GNS [40]. However, its tumor uptake behavior, pharmacokinetics, biodistribution, in vivo photothermal therapy, long-term toxicity have not yet been investigated. This study is aimed to investigate those crucial GNS properties for future clinical translation.

1.3 Surface Enhanced Raman Spectroscopy (SERS)

Surface-enhanced Raman spectroscopy (SERS) takes advantage of the surface plasmon resonance to amplify Raman scattering, resulting in a sensitive and specific method for chemical analysis [41, 42]. Raman spectroscopy is a non-destructive photon-scattering technique, which provides useful spectral information related to specific molecular vibrational energy levels of the analyzed molecules [10, 41, 43, 44]. However, the intrinsic Raman scattering efficiency is low: even strong Raman scatter exhibits cross-sections only on the order of 10^{-29} cm^2 $molecule^{-1}$ sr^{-1} while the fluorescence cross-section is usually on the order of 10^{-16} cm^2 $molecule^{-1}$ sr^{-1} [45]. The SERS effect was discovered in the mid-1970s [8, 9] and the first analytical application of SERS in chemical analysis was demonstrated in our laboratory [46–48]. Extensive research has been devoted to understanding and modeling the Raman enhancement in SERS since the mid-1980s and results suggested that the SERS effect primarily arises from electromagnetic resonance occurring near metallic nanostructure surface [49–53]. When the metallic nanostructure surface is irradiated by an external electromagnetic field, electrons within the conduction band oscillate at the same frequency. These oscillating electrons are known as surface plasmons, which produce a secondary electromagnetic field added to the external electromagnetic field, resulting in the surface plasmon resonance [42, 51]. The secondary electric field is typically most concentrated on the rough metallic nanostructure surface. In addition, the SERS effect can also be due to a chemical enhancement associated with interaction between molecules and metal surface [54]. The resulting SERS enhancement factor can be 10^6 to 10^8-fold, and up to 10^{15}-fold at "hot spots" where electromagnetic field enhancements of multiple plasmonic nanoparticles overlap in a small space [45, 55]. With such a high sensitivity, SERS has been shown to be a powerful chemical analysis technique with sensitivity reaching the single-molecule level [56].

1.4 Two-Photon Photoluminescence (TPL)

Optical imaging has been widely used for biomedical applications including preclinical study and intraoperative tumor margin delineation [15, 57–59]. It has advantages of low cost, high spatial and temporal resolution [60, 61]. However, traditional optical imaging like single photon fluorescence suffers from low tissue penetration depth [62]. Two-photon photoluminescence (TPL) provides an improved modality with deeper tissue imaging capability [63]. TPL typically applies near-infrared (NIR) light as excitation source, which is considered as the "tissue optical window" because NIR light with low absorption and scattering has deeper tissue penetration [19]. For TPL, two excitation photons are absorbed simultaneously to excite one electron from an occupied band to an unoccupied band and then the excited electron could be moved to be closer to the Fermi level with intraband scattering. After that, one luminescence photon can be released by the electron-pair relaxation process. The excitation light has longer wavelength than the emission light and the background noise, which is usually found in single photon microscopy, can be largely reduced. TPL has quadratic dependence on the excitation intensity and is highly sensitive to the local surface plasmon resonances (LSPR), which is similar as SERS [64].

The GNS with tunable plasmonics in the NIR region provides a superior TPL contrast agent with high intrinsic signal to investigate nanoparticle pharmacokinetics, biodistribution and cell uptake mechanisms in real time. GNS has significantly higher TPL cross section than organic dyes and other gold nanoparticles and it has been reported that GNS's TPL cross section is several orders of magnitude higher than other contrast agents [22]. The intrinsic high TPL signal enables GNS to be visualized without the need of extra dye labeling, making their imaging applications more convenient and free of potential dye leaching problems.

1.5 X-ray Computed Tomography (CT)

X-ray computed tomography (CT) reconstructs a large number of X-ray images taken from various angles to generate tomographic images of the scanned object [65, 66]. It is a fast and powerful anatomical imaging method and widely used in clinical imaging and preclinical studies. CT is very useful for tumor detection and treatment planning for both surgery and radiation therapy [67–69]. However, its applications are limited by low contrast discrimination and usually require high concentration of contrast agent administration. The atomic number of gold (79) is much higher than that of clinically used iodine, which is clinically used for CT imaging (53). Therefore, GNS can induce stronger X-ray attenuation and have potential to be used as contrast agent for CT scans [33]. The detection sensitivity to gold can be enhanced by using dual energy (DE-CT) CT, which applies 2 acquisitions with different X-ray energies that are used to distinguish imaging targets with various attenuation properties [70–73]. In addition, DE-CT imaging provides non-invasive real time dynamic information, which allows accurately determining GNS concentrations in different tissues at any given time [73–75]. This allows acquiring biodistribution results dynamically, without having to sacrifice animals at each time point. It also allows visualizing well-vascularized tumor regions for nanoparticle delivery through EPR effect.

1.6 Magnetic Resonance Imaging (MRI)

Magnetic resonance imaging (MRI) is an important modality for biomedical imaging [76–81]. It has no ionizing radiation and been considered to be superior to CT imaging despite of higher cost and longer scan time [80, 82, 83]. MRI technique is based on the nuclear magnetic resonance (NMR) and ultra high magnetic field, radio waves and field gradients are used to form 3D images with resonance signal encoded in each dimensions [78, 81, 84–86]. In preclinical and clinical applications, hydrogen atoms are typically used to generate signal and as a result, the MRI is basically to show the tissue distribution of water and fat, which contain high amount of hydrogen atoms [81, 87]. The voxel intensity of the tissue sample is determined by the proton density and tissue-specific relaxation time. For molecular imaging, the MRI's sensitivity is typically in the range of 10^{-3} to 10^{-5} mol/L [88]. MRI sensitivity can be improved by increasing magnetic field strength but the instrument cost can be dramatically increased. Gadolinium chelates have been widely used to enhance MRI signal and the typical dose is 100 μmol/kg [89, 90]. Targeted MRI contrast agents have been developed by linking contrast agents to targeting ligands such as antibody for in vivo cancer imaging [37, 85, 91]. GNS can also be linked with contrast agent, such as Gd, by using DOTA chelator for MRI imaging.

1.7 Positron Emission Tomography (PET)

Positron emission tomography (PET) is a noninvasive imaging technique for sensitive radiotracer detection in 3D and has been used for cancer imaging in clinical applications [92–98]. In a positron emission process (positive beta decay), a proton of a radioisotope is converted to a neutron and produces a positron (e^+, an antiparticle of the electron with opposite charge) and an electron neutrino [99–101].

The generated positron travels a short distance and interacts with an electron to get annihilated, generating a pair of gamma photons emitting at almost 180 degrees to each other. The gamma photon can create a light burst when they encounter a scintillator. The generated light can be detected by photomultiplier tubes or other light detectors. A real event is recognized only when the simultaneous detection of a pair of gamma photons in opposite directions happens [62]. Photons are ignored if they are not detected in pairs. Therefore, the detection background noise is significantly suppressed. After measurement, coincidence events can be analyzed and reconstructed to get the 3D distribution of the radiotracer [102]. There are several radioisotopes used for PET imaging, including ^{18}F, ^{64}Cu and ^{124}I [4, 95, 103–106]. Fluorine-18 fluorodeoxyglucose (FDG) has been widely used in clinical oncology for cancer metastasis detection because cancer has been found to have higher glucose uptake than surrounding normal tissues [96, 97, 107, 108]. PET imaging is often combined with CT to show radiotracer distribution in anatomical structure [94, 109, 110]. Currently, PET scan has also been combined with MRI, which has better discrimination capability for soft tissue. In addition to oncology applications, the noninvasive PET 3D imaging is also a powerful tool to investigate drug pharmacokinetics [111]. The tissue uptake and biodistribution can be dynamically monitored with PET scan at different time points without need of samples collection or sacrificing animals. GNS nanoprobe can be radiolabeled with ^{64}Cu and ^{124}I for PET imaging to be used for in vivo tracking and sensitive cancer detection.

1.8 Photoimmunotherapy

Photoimmunotherapy, a combination of photothermal therapy and immunotherapy, has the potential to treat not only primary tumor with specific thermal ablation but also cancer metastasis with triggered immune response. Activating the immune system for therapeutic benefit in cancer has long been a goal in immunology and oncology [112–117]. Photothermal therapy (PTT) exploiting high temperature transduced by plasmonic nanoparticles from photon energy is a new promising method offering efficient and specific cancer treatment [118–121]. Historically, thermal therapy has been used to treat cancers clinically because cancer cells are more sensitive to elevated temperature than normal cells [122–127]. With targeting GNS accumulating in the tumor, PTT brings potential for delivering ablation more precisely [24]. Immunotherapies have emerged as one of the most-promising modalities to treat cancer and have involved the use of drugs that target immune checkpoints [128]. In recent years, immunotherapy with specific immune checkpoint inhibitor provides a promising way to break tumor immunosuppressive environment [129]. Programmed death-ligand 1 (PD-L1), a protein overexpressed on cancer cell membrane, contributes to the suppression of the immune system [129]. PD-L1 binds to its receptor, programmed death protein 1 (PD-1), found on activated T cells, B cells, and myeloid cells, to modulate T cell function [130]. The therapeutic anti-PD-L1 antibody is designed to block the PD-1/PD-L1 interaction and reverse tumor-mediated immunosuppression [112, 128, 131, 132]. Immune blockade inhibitors targeting PD-1 or its ligand PD-L1, have emerged as one of the most-promising immune therapy modalities to treat systemic cancer [133, 134]. But currently, they work only for a limited number of patients and can become ineffective with time. There is a strong need to combine other complementary therapies with PD-1/PD-L1 immunotherapy to improve therapeutic response in treating cancer metastasis.

It has been reported that damaged cancer cells after thermal treatment exposes tumor-specific antigens to antigen-presenting cells (APCs) with the help of heat shock proteins, resulting in the generation of immune response to kill those cancer cells in both primary tumor location as well as metastasis regions [135–140]. Therefore, a potential synergistic effect between immunotherapy and photothermal therapy would be expected to improve cancer therapy. Broadening and stabilizing the effect of PD-1/PD-L1 inhibitors with plasmonic nanoparticle-mediated photothermal therapy could potentially address the challenge of metastatic cancer treatment.

1.9 Summary and Study Outline

This study is aimed at developing and optimizing GNS multifunctional nanoprobes for in vivo sensitive cancer detection and photoimmunotherapy. We started from multifunctional GNS nanoprobe development and characterization (Chap. 2). Sensitivity of different imaging modalities was compared and the photon-to-heat conversion efficiency of GNS nanoprobe was measured in solution. The therapeutic or diagnostic radioisotope loading capability was also demonstrated. Next, in vivo evaluation and optimization were performed (Chap. 3). Effects of nanoparticle size, surface coating, injection dose, and circulation time after intravenous (IV) injection, on GNS tumor uptake were studied. The GNS toxicity was also evaluated with both in vitro cell study and long term in vivo study with the mouse animal model (Chap. 4). Body weight monitoring, blood chemistry test and histopathology examination were performed to investigate potential deleterious effects after GNS IV administration up to 6 months. After optimization, the PEGylated 30 nm GNS nanoprobe with ^{124}I radioisotope labeling was chosen for sensitive brain tumor detection down to 0.5 mm using noninvasive PET imaging (Chap. 5). The GNS nanoprobe was demonstrated to be much better than current clinically used contrast agent, fludeoxyglucose (^{18}F) (FDG), for brain tumor detection. The transmission electron microscopy (TEM) was performed on brain tumor after GNS IV infusion. GNS nanoprobes were found to be endocytosized inside brain tumor cells after leaking through tumor-disrupted blood-brain-barrier. The GNS-enhanced photoimmnotherapy was performed with bladder cancer mouse model for both primary tumor and cancer metastasis treatment (Chap. 6). Therapeutic response was observed not only for the primary tumor with photothermal ablation but also the distant tumor (to mimic cancer metastasis) without photothermal treatment. It should be noted that a portion of mice had both tumors eradicated for 4 months (complete cure) and had no tumor developed after cancer cells injection (rechallenge test), indicating the generation of memorized immune response like cancer vaccine effect. A conclusion and future outlook will be discussed in the Chap. 7. The biocompatible multifunctional GNS nanoprobe with superior capabilities for cancer detection and treatment shows great potential for future clinical translation in cancer management.

References

1. Ferlay J, Shin HR, Bray F, Forman D, Mathers C, Parkin DM (2010) Estimates of world wide burden of cancer in 2008: GLOBOCAN 2008. Int J Cancer 127:2893–2917
2. Unger-Saldaña K (2014) Challenges to the early diagnosis and treatment of breast cancer in developing countries. World J Clin Oncol 5:465–477
3. Cheng Y, Morshed RA, Auffinger B, Tobias AL, Lesniak MS (2014) Multifunctional nanoparticles for brain tumor imaging and therapy. Adv Drug Deliv Rev 66:42–57
4. Wang L, Wang Y, Li Z (2013) Nanoparticle-based tumor theranostics with molecular imaging. Curr Pharm Biotechnol 14:683–692
5. Zhao L, Kim T-H, Ahn J-C, Kim H-W, Kim SY (2013) Highly efficient "theranostics" system based on surface-modified gold nanocarriers for imaging and photodynamic therapy of cancer. J Mat Chem B 1:5806–5817
6. Mieszawska AJ, Mulder WJM, Fayad ZA, Cormode DP (2013) Multifunctional gold nanoparticles for diagnosis and therapy of disease. Mol Pharm 10:831–847
7. Yuan H, Khoury CG, Hwang H, Wilson CM, Grant GA, Vo-Dinh T (2012) Gold nanostars: surfactant-free synthesis, 3D modelling, and two-photon photoluminescence imaging. Nanotechnology 23(075102):1–9
8. Yuan H, Fales AM, Vo-Dinh T (2012) TAT peptide-functionalized gold nanostars: enhanced intracellular delivery and efficient NIR photothermal therapy using ultralow irradiance. J Am Chem Soc 134:11358–11361
9. Sun XL et al (2014) Chelator-free Cu-64-integrated gold nanomaterials for positron emission tomography imaging guided photothermal cancer therapy. ACS Nano 8:8438–8446
10. Bayazitoglu Y, Kheradmand S, Tullius TK (2013) An overview of nanoparticle assisted laser therapy. Int J Heat Mass Transf 67:469–486
11. Ma Y, Liang X, Tong S, Bao G, Ren Q, Dai Z (2013) Gold nanoshell nanomicelles for potential magnetic resonance imaging, light-triggered drug release, and photothermal therapy. Adv Func Mater 23:815–822
12. Khlebtsov N, Dykman L (2011) Biodistribution and toxicity of engineered gold nanoparticles: a review of in vitro and in vivo studies. Chem Soc Rev 40:1647–1671
13. Cole JR, Mirin NA, Knight MW, Goodrich GP, Halas NJ (2009) Photothermal efficiencies of nanoshells and nanorods for clinical therapeutic applications. J Phys Chem C 113:12090–12094
14. Zhang Z, Wang J, Chen C (2013) Gold nanorods based platforms for light-mediated theranostics. Theranostics 3:223–238
15. Zhang Y, Qian J, Wang D, Wang Y, He S (2013) Multifunctional gold nanorods with ultrahigh stability and tunability for in vivo fluorescence imaging, SERS detection, and photodynamic therapy. Angewandte Chemie-Int Ed 52:1148–1151
16. Agarwal A et al (2011) Dual-mode imaging with radiolabeled gold nanorods. J Biomed Opt 16(051307):1–7

17. Tong L, Wei Q, Wei A, Cheng J-X (2009) Gold nanorods as contrast agents for biological imaging: optical properties, surface conjugation and photothermal effects. Photochem Photobiol 85:21–32
18. Chen JY et al (2007) Immuno gold nanocages with tailored optical properties for targeted photothermal destruction of cancer cells. Nano Lett 7:1318–1322
19. Weissleder R (2001) A clearer vision for in vivo imaging. Nat Biotechnol 19:316–317
20. Pekkanen AM, DeWitt MR, Rylander MN (2014) Nanoparticle enhanced optical imaging and phototherapy of cancer. J Biomed Nanotechnol 10:1677–1712
21. Schlucker S (2014) Surface-enhanced Raman spectroscopy: concepts and chemical applications. Angewandte Chemie-Int Ed 53:4756–4795
22. Gao N et al (2014) Shape-dependent two-photon photoluminescence of single gold nanoparticles. J Phys Chem C 118:13904–13911
23. Fales AM, Yuan H, Vo-Dinh T (2011) Silica-coated gold nanostars for combined surface-enhanced Raman scattering (SERS) detection and singlet-oxygen generation: a Potential nanoplatform for theranostics. Langmuir 27:12186–12190
24. Liu Y et al (2015) A plasmonic gold nanostar theranostic probe for in vivo tumor imaging and photothermal therapy. Theranostics 5:946–960
25. Day ES et al (2011) Nanoshell-mediated photothermal therapy improves survival in a murine glioma model. J Neuro-Oncol 104:55–63
26. Pattani VP, Tunnell JW (2012) Nanoparticle-mediated photothermal therapy: a comparative study of heating for different particle types. Lasers Surg Med 44:675–684
27. Kumar D, Saini N, Jain N, Sareen R, Pandit V (2013) Gold nanoparticles: an era in bionanotechnology. Expert Opin Drug Deliv 10:397–409
28. Lee HY et al (2008) PET/MRI dual-modality tumor imaging using arginine-glycine-aspartic (RGD)—conjugated radiolabeled iron oxide nanoparticles. J Nucl Med 49:1371–1379
29. Maeda H, Wu J, Sawa T, Matsumura Y, Hori K (2000) Tumor vascular permeability and the EPR effect in macromolecular therapeutics: a review. J Control Release 65:271–284
30. Fang J, Nakamura H, Maeda H (2011) The EPR effect: unique features of tumor blood vessels for drug delivery, factors involved, and limitations and augmentation of the effect. Adv Drug Deliv Rev 63:136–151
31. Gabizon A, Shmeeda H, Barenholz Y (2003) Pharmacokinetics of pegylated liposomal doxorubicin—Review of animal and human studies. Clin Pharmacokinet 42:419–436
32. Brigger I, Dubernet C, Couvreur P (2002) Nanoparticles in cancer therapy and diagnosis. Adv Drug Deliv Rev 54:631–651
33. Lee N, Choi SH, Hyeon T (2013) Nano-sized CT contrast agents. Adv Mater 25:2641–2660
34. Xing Y, Zhao J, Conti PS, Chen K (2014) Radiolabeled nanoparticles for multimodality tumor imaging. Theranostics 4:290–306
35. Perez-Campana C et al (2013) Biodistribution of different sized nanoparticles assessed by positron emission tomography: a general strategy for direct activation of metal oxide particles. ACS Nano 7:3498–3505
36. Kircher MF et al (2012) A brain tumor molecular imaging strategy using a new triple-modality MRI-photoacoustic-Raman nanoparticle. Nat Med 18:829–834
37. Gerion D et al (2007) Paramagnetic silica-coated nanocrystals as an advanced MRI contrast agent. J Phys Chem C 111:12542–12551
38. Yuan H, Register JK, Wang HN, Fales AM, Liu Y, Vo-Dinh T (2013) Plasmonic nanoprobes for intracellular sensing and imaging. Anal Bioanal Chem 405:6165–6180
39. Gad SC, Sharp KL, Montgomery C, Payne JD, Goodrich GP (2012) Evaluation of the toxicity of intravenous delivery of Auroshell particles (gold-silica nanoshells). Int J Toxicol 31:584–594
40. Yuan H, Khoury CG, Hwang H, Wilson CM, Grant GA, Vo-Dinh T (2012) Gold nanostars: surfactant-free synthesis, 3D modelling, and two-photon photoluminescence imaging. Nanotechnology 23(075102):1–7

41. Driscoll AJ, Harpster MH, Johnson PA (2013) The development of surface-enhanced Raman scattering as a detection modality for portable in vitro diagnostics: progress and challenges. Phys Chem Chem Phys 15:20415-20433
42. Kerker M (1984) Electromagnetic model for surface-enhanced Raman scattering (SERS) on metal colloids. Acc Chem Res 17:271–277
43. Cialla D et al (2012) Surface-enhanced Raman spectroscopy (SERS): progress and trends. Anal Bioanal Chem 403:27–54
44. Nafie LA (2011) Recent advances in linear and nonlinear Raman spectroscopy. Part V. J Raman Spectrosc 42:2049–2068
45. Scaffidi JP, Gregas MK, Seewaldt V, Vo-Dinh T (2009) SERS-based plasmonic nanobiosensing in single living cells. Anal Bioanal Chem 393:1135–1141
46. Albrecht MG, Creighton JA (1977) Anomalously intense Raman-spectra of pyridine at a silver electrode. J Am Chem Soc 99:5215–5217
47. Fleischmann M, Hendra PJ (1974) McQuilla.Aj. Raman-spectra of pyridine adsorbed at a silver electrode. Chem Phys Lett 26:163–166
48. Vo-Dinh T, Hiromoto MYK, Begun GM, Moody RL (1984) Surface-enhanced Raman spectrometry for trace organic analysis. Anal Chem 56:1667–1670
49. Graham D, Smith WE, Linacre AMT, Munro CH, Watson ND, White PC (1997) Selective detection of deoxyribonucleic acid at ultralow concentrations by SERRS. Anal Chem 69:4703–4707
50. Kneipp K et al (1997) Single molecule detection using surface-enhanced Raman scattering (SERS). Phys Rev Lett 78:1667–1670
51. Vo-Dinh T (1998) Surface-enhanced Raman spectroscopy using metallic nanostructures. Trac-Trends in Anal Chem 17:557–582
52. Schatz GC (1984) Theoretical studies of surface enhanced Raman scattering. Acc Chem Res 17:370–376
53. Moskovits M (1985) Surface-enhanced spectroscopy. Rev Mod Phys 57:783–826
54. Tuan V-D et al (2013) Plasmonic nanoprobes: from chemical sensing to medical diagnostics and therapy. Nanoscale 5:10127–10140
55. Liu Y, Chang Z, Yuan H, Fales AM, Vo-Dinh T (2013) Quintuple-modality (SERS-MRI-CT-TPL-PTT) plasmonic nanoprobe for theranostics. Nanoscale 5:12126–12131
56. Nie SM, Emery SR (1997) Probing single molecules and single nanoparticles by surface-enhanced Raman scattering. Science 275:1102–1106
57. Kagadis GC, Loudos G, Katsanos K, Langer SG, Nikiforidis GC (2010) In vivo small animal imaging: current status and future prospects. Med Phys 37:6421–6442
58. Yan H et al (2012) Two-order targeted brain tumor imaging by using an optical/paramagnetic nanoprobe across the blood brain barrier. ACS Nano 6:410–420
59. Whitley MJ et al (2016) A mouse-human phase 1 co-clinical trial of a protease-activated fluorescent probe for imaging cancer. Sci Transl Med 8:11
60. Hilderbrand SA, Weissleder R (2010) Near-infrared fluorescence: application to in vivo molecular imaging. Curr Opin Chem Biol 14:71–79
61. Vahrmeijer AL, Hutteman M, van der Vorst JR, van de Velde CJH, Frangioni JV (2013) Image-guided cancer surgery using near-infrared fluorescence. Nat Rev Clin Oncol 10: 507–518
62. Koo V, Hamilton PW, Williamson K (2006) Non-invasive in vivo imaging in small animal research. Cell Oncol 28:127–139
63. Helmchen F, Denk W (2005) Deep tissue two-photon microscopy. Nat Methods 2:932–940
64. Bouhelier A, Bachelot R, Lerondel G, Kostcheev S, Royer P, Wiederrecht GP (2005) Surface plasmon characteristics of tunable photoluminescence in single gold nanorods. Phys Rev Lett 95(267405):1–4
65. Betzer O et al (2014) Nanoparticle-based CT imaging technique for longitudinal and quantitative stem cell tracking within the brain: application in neuropsychiatric disorders. ACS Nano 8:9274–9285

66. Lederlin M et al (2012) In vivo micro-CT assessment of airway remodeling in a flexible OVA-sensitized murine model of asthma. PLoS ONE 7(e48493):1–12
67. US preventive services task force recommendation statement (2014) Moyer VA, Force USPST. Screening for lung cancer. Ann Intern Med 160:330–338
68. Bach PB et al (2012) Benefits and harms of CT screening for lung cancer: a systematic review. JAMA-J Am Med Assoc 307:2418–2429
69. Swensen SJ et al (2005) CT screening for lung cancer: five-year prospective experience. Radiology 235:259–265
70. Badea CT, Johnston S, Johnson B, Lin M, Hedlund LW, Johnson GA (2008) A dual micro-CT system for small animal imaging—art. no. 691342. In: Hsieh J, Samei E (eds) Medical imaging 2008: physics of medical imaging, Pts 1–3
71. Graser A, Johnson TRC, Chandarana H, Macari M (2009) Dual energy CT: preliminary observations and potential clinical applications in the abdomen. Eur Radiol 19:13–23
72. Petersilka M, Bruder H, Krauss B, Stierstorfer K, Flohr TG (2008) Technical principles of dual source CT. Eur J Radiol 68:362–368
73. Johnson TRC et al (2007) Material differentiation by dual energy CT: initial experience. Eur Radiol 17:1510–1517
74. Ashton JR et al (2014) Dual-energy micro-CT functional imaging of primary lung cancer in mice using gold and iodine nanoparticle contrast agents: a validation study. PLoS ONE 9 (e88129):1–14
75. Wang ZM et al (1996) Skeletal muscle mass: evaluation of neutron activation and dual-energy X-ray absorptiometry methods. J Appl Physiol 80:824–831
76. Houssami N et al (2008) Accuracy and surgical impact of magnetic resonance imaging in breast cancer staging: systematic review and meta-analysis in detection of multifocal and multicentric cancer. J Clin Oncol 26:3248–3258
77. Mann RM, Hoogeveen YL, Blickman JG, Boetes C (2008) MRI compared to conventional diagnostic work-up in the detection and evaluation of invasive lobular carcinoma of the breast: a review of existing literature. Breast Cancer Res Treat 107:1–14
78. Sun C, Lee JSH, Zhang MQ (2008) Magnetic nanoparticles in MR imaging and drug delivery. Adv Drug Deliv Rev 60:1252–1265
79. Koh DM, Collins DJ (2007) Diffusion-weighted MRI in the body: Applications and challenges in oncology. Am J Roentgenol 188:1622–1635
80. Kriege M et al (2004) Efficacy of MRI and mammography for breast-cancer screening in women with a familial or genetic predisposition. N Engl J Med 351:427–437
81. Kobayashi H, Kawamoto S, Jo SK, Bryant HL, Brechbiel MW, Star RA (2003) Macromolecular MRI contrast agents with small dendrimers: pharmacokinetic differences between sizes and cores. Bioconjug Chem 14:388–394
82. Bipat S, Glas AS, van der Velden J, Zwinderman AH, Bossuyt PMM, Stoker J (2003) Computed tomography and magnetic resonance imaging in staging of uterine cervical carcinoma: a systematic review. Gynecol Oncol 91:59–66
83. Saslow D et al (2007) American cancer society guidelines for breast screening with MRI as an adjunct to mammography. Ca-Cancer J Clin 57:75–89
84. Moffat BA, Galban CJ, Rehemtulla A (2009) Advanced MRI: translation from animal to human in brain tumor research. Neuroimaging Clin N Am 19:517–526
85. Lipinski MJ et al (2006) MRI to detect atherosclerosis with gadolinium-containing immunomicelles targeting the macrophage scavenger receptor. Magn Reson Med 56: 601–610
86. Tofts PS et al (1999) Estimating kinetic parameters from dynamic contrast-enhanced T(1)-weighted MRI of a diffusable tracer: standardized quantities and symbols. J Magn Reson Imaging 10:223–232
87. Eisenhauer EA et al (2009) New response evaluation criteria in solid tumours: revised RECIST guideline (version 1.1). Eur J Cancer 45:228–247
88. Caravan P (2006) Strategies for increasing the sensitivity of gadolinium based MRI contrast agents. Chem Soc Rev 35:512–523

89. Strijkers GJ, Mulder WJM, van Tilborg GAF, Nicolay K (2007) MRI contrast agents: current status and future perspectives. Anti-Cancer Agents Med Chem 7:291–305
90. Anzalone N et al (2013) Cerebral neoplastic enhancing lesions: multicenter, randomized, crossover intraindividual comparison between gadobutrol (1.0 M) and gadoterate meglumine (0.5 M) at 0.1 mmol Gd/kg body weight in a clinical setting. Eur J Radiol 82:139–145
91. Sipkins DA, Cheresh DA, Kazemi MR, Nevin LM, Bednarski MD, Li KCP (1998) Detection of tumor angiogenesis in vivo by alpha(v)beta(3)-targeted magnetic resonance imaging. Nat Med 4:623–626
92. O'Connor JPB, Jackson A, Asselin MC, Buckley DL, Parker GJM, Jayson GC (2008) Quantitative imaging biomarkers in the clinical development of targeted therapeutics: current and future perspectives. Lancet Oncol 9:766–776
93. Fletcher JW et al (2008) Recommendations on the use of F-18-FDG PET in oncology. J Nucl Med 49:480–508
94. Antoch G et al (2003) Whole-body dual-modality PET/CT and whole-body MRI for tumor staging in oncology. JAMA-J Am Med Assoc 290:3199–3206
95. Zhuang HM et al (2001) Dual time point F-18-FDG PET imaging for differentiating malignant from inflammatory processes. J Nucl Med 42:1412–1417
96. Vansteenkiste JF et al (1999) Prognostic importance of the standardized uptake value on F-18-fluoro-2-deoxy-glucose-positron emission tomography scan in non-small-cell lung cancer: an analysis of 125 cases. J Clin Oncol 17:3201–3206
97. Shreve PD, Anzai Y, Wahl RL (1999) Pitfalls in oncologic diagnosis with FDG PET imaging: physiologic and benign variants. Radiographics 19:61–77
98. Shields AF et al (1998) Imaging proliferation in vivo with F-18 FLT and positron emission tomography. Nat Med 4:1334–1336
99. Rohren EM, Turkington TG, Coleman RE (2004) Clinical applications of PET in oncology. Radiology 231:305–332
100. Gambhir SS (2002) Molecular imaging of cancer with positron emission tomography. Nat Rev Cancer 2:683–693
101. Beyer T et al (2000) A combined PET/CT scanner for clinical oncology. J Nucl Med 41:1369–1379
102. Phelps ME (2000) PET: the merging of biology and imaging into molecular imaging. J Nucl Med 41:661–681
103. Pal A et al (2006) Molecular imaging of EGFR kinase activity in tumors with I-124-labeled small molecular tracer and positron emission tomography. Mol Imag Biol 8:262–277
104. Sgouros G et al (2004) Patient-specific dosimetry for I-131 thyroid cancer therapy using I-124 PET and 3-dimensional-internal dosimetry (3D-ID) software. J Nucl Med 45: 1366–1372
105. Sundaresan G et al (2003) I-124-labeled engineered Anti-CEA minibodies and diabodies allow high-contrast, antigen-specific small-animal PET imaging of xenografts in athymic mice. J Nucl Med 44:1962–1969
106. Pentlow KS et al (1996) Quantitative imaging of iodine-124 with PET. J Nucl Med 37:1557–1562
107. Schoder H, Yeung HWD, Gonen M, Kraus D, Larson SM (2004) Head and neck cancer: clinical usefulness and accuracy of PET/CT image fusion. Radiology 231:65–72
108. Bradley J et al (2004) Impact of FDG-PET on radiation therapy volume delineation in non-small-cell lung cancer. Int J Radiat Oncol Biol Phys 59:78–86
109. Boellaard R et al (2010) FDG PET and PET/CT: EANM procedure guidelines for tumour PET imaging: version 1.0. Eur J Nucl Med Mol Imaging 37:181–200
110. Antoch G et al (2003) Non-small cell lung cancer: dual-modality PET/CT in preoperative staging. Radiology 229:526–533
111. Haubner R et al (2001) Glycosylated RGD-containing peptides, tracer for tumor targeting and angiogenesis imaging with improved biokinetics. J Nucl Med 42:326–336

112. Gandini S, Massi D, Mandala M (2016) PD-L1 expression in cancer patients receiving anti PD-1/PD-L1 antibodies: a systematic review and meta-analysis. Crit Rev Oncol Hematol 100:88–98
113. Niezgoda A, Niezgoda P, Czajkowski R (2015) Novel approaches to treatment of advanced melanoma: a review on targeted therapy and immunotherapy. Biomed Res Int 2015 (851387):1–16
114. Melero I, Berman DM, Aznar MA, Korman AJ, Gracia JLP, Haanen J (2015) Evolving synergistic combinations of targeted immunotherapies to combat cancer. Nat Rev Cancer 15:457–472
115. Mamalis A, Garcha M, Jagdeo J (2014) Targeting the PD-1 pathway: a promising future for the treatment of melanoma. Arch Dermatol Res 306:511–519
116. Simeone E, Ascierto PA (2012) Immunomodulating antibodies in the treatment of metastatic melanoma: the experience with anti-CTLA-4, anti-CD137, and anti-PD1. J Immunotoxicol 9:241–247
117. Alexandrescu DT et al (2010) Immunotherapy for melanoma: current status and perspectives. J Immunother 33:570–590
118. Lal S, Clare SE, Halas NJ (2008) Nanoshell-enabled photothermal cancer therapy: impending clinical impact. Acc Chem Res 41:1842–1851
119. El-Sayed IH, Huang XH, El-Sayed MA (2006) Selective laser photo-thermal therapy of epithelial carcinoma using anti-EGFR antibody conjugated gold nanoparticles. Cancer Lett 239:129–135
120. O'Neal DP, Hirsch LR, Halas NJ, Payne JD, West JL (2004) Photo-thermal tumor ablation in mice using near infrared-absorbing nanoparticles. Cancer Lett 209:171–176
121. Hirsch LR et al (2003) Nanoshell-mediated near-infrared thermal therapy of tumors under magnetic resonance guidance. Proc Natl Acad Sci U S A 100:13549–13554
122. Davalos RV, Mir LM, Rubinsky B (2005) Tissue ablation with irreversible electroporation. Ann Biomed Eng 33:223–231
123. Abdalla EK et al (2004) Recurrence and outcomes following hepatic resection, radiofrequency ablation, and combined resection/ablation for colorectal liver metastases. Ann Surg 239:818–824
124. Goldberg SN, Gazelle GS, Mueller PR (2000) Thermal ablation therapy for focal malignancy: a unified approach to underlying principles, techniques, and diagnostic imaging guidance. Am J Roentgenol 174:323–331
125. Patterson EJ, Scudamore CH, Nagy AG, Buczkowski AK (1998) Radiofrequency ablation of porcine liver in vivo—effects of blood flow and treatment time on lesion size. Ann Surg 227:559–565
126. Lin SM, Lin CJ, Lin CC, Hsu CW, Chen YC (2005) Randomised controlled trial comparing percutaneous radiofrequency thermal ablation, percutaneous ethanol injection, and percutaneous acetic acid injection to treat hepatocellular carcinoma of 3 cm or less. Gut 54: 1151–1156
127. Goldberg SN et al (2005) Image-guided tumor ablation: standardization of terminology and reporting criteria. Radiology 235:728–739
128. Nowicki TS, Anderson JL, Federman N (2016) Prospective immunotherapies in childhood sarcomas: PD1/PDL1 blockade in combination with tumor vaccines. Pediatr Res 79: 371–377
129. Merelli B, Massi D, Cattaneo L, Mandala M (2014) Targeting the PD1/PD-L1 axis in melanoma: biological rationale, clinical challenges and opportunities. Crit Rev Oncol Hematol 89:140–165
130. Chen DS, Mellman I (2013) Oncology meets immunology: the cancer-immunity cycle. Immunity 39:1–10
131. Xia Y, Medeiros LJ, Young KH (2016) Signaling pathway and dysregulation of PD1 and its ligands in lymphoid malignancies. Biochimica Et Biophysica Acta-Reviews on Cancer 1865:58–71

132. Iacovelli R et al (2016) Prognostic role of PD-L1 expression in renal cell carcinoma. A systematic review and meta-analysis. Target Oncol 11:143–148
133. Seyedin SN, Tang C, Welsh JW (2015) Author's view: radiation and immunotherapy as systemic therapy for solid tumors. Oncoimmunology 4(e986402):1–3
134. Ruiz R, Hunis B, Raez LE (2014) Immunotherapeutic agents in non-small-cell lung cancer finally coming to the front lines. Curr Oncol Rep 16(400):1–10
135. Toraya-Brown S et al (2014) Local hyperthermia treatment of tumors induces CD8(+) T cell-mediated resistance against distal and secondary tumors. Nanomed-Nanotechnol Biol Med 10:1273–1285
136. Toraya-Brown S, Fiering S (2014) Local tumour hyperthermia as immunotherapy for metastatic cancer. Int J Hyperth 30:531–539
137. Du GJ, Liu YH, Li JH, Liu WJ, Wang YY, Li H (2013) Hypothermic microenvironment plays a key role in tumor immune subversion. Int Immunopharmacol 17:245–253
138. Kubes J, Svoboda J, Rosina J, Starec M, Fiserova A (2008) Immunological response in the mouse melanoma model after local hyperthermia. Physiol Res 57:459–465
139. Kida Y et al (2006) Increased liver temperature efficiently augments human cellular immune response: T-cell activation and possible monocyte translocation. Cancer Immunol Immunother 55:1459–1469
140. Milani V et al (2002) Heat shock protein 70: role in antigen presentation, and immune stimulation. Int J Hyperth 18:563–575

Chapter 2
Multifunctional GNS Nanoprobe Development and Characterization

2.1 Multifunctional GNS Nanoprobe Development

The developed multifunctional GNS nanoprobe is illustrated in Fig. 2.1. The GNS nanoparticles were synthesized by using a seed-mediated and surfactant-free method [1, 2]. First, 12 nm gold nanosphere seeds were produced via the reduction of $HAuCl_4$ (100 ml, 1 mM) by trisodium citrate (15 ml, 1% w/v) at boiling temperature in water solution. After that, GNS nanoparticles were synthesized by rapidly and simultaneously mixing $AgNO_3$ (1 ml, 3 mM) and ascorbic acid (500 μl, 0.1 M) with 100 ml of solution containing 0.25 mM $HAuCl_4$, 1 mM HCl, and 1 ml of the 12 nm gold nanosphere seeds. The GNS nanoparticle size can be tuned by changing the concentration ratio between $HAuCl_4$ and 12 nm gold nanosphere seeds. After synthesis, 2 μM (final concentration) of para-mercaptobenzoic acid (pMBA), a Raman reporter, was added into GNS solution and incubated for 10 min followed by adding 2 μM (final concentration) mPEG-SH (MW 5000) for 1 h incubation. Following that, a thin layer of silica was coated on the GNS by reaction with tetraethyl orthosilicate (TEOS). After thiol-functionalization with 3-mercaptopropyltrimethoxysilane (MPTMS) on the coated silica surface, maleimido-mono-amide-DOTA (1,4,7,10-tetraazacyclododecane-1,4,7,10-tetraacetic acid) (Macrocylics, TX) was added at a molar excess ratio of 2×10^6 per nanoparticle from stock. After 8 h of reaction at room temperature, synthesized GNS nanoparticles with DOTA chelator were centrifuge washed (5000 g for 10 min) three times for purification. The gadolinium chloride was then added to the DOTA-functionalized nanoparticles with a Gd^{3+} to nanoparticle ratio of 2×10^6, in 10 mM 2-(*N*-morpholino) ethanesulfonic acid (MES) buffer at pH 6.3 with 0.1% sodium dodecyl sulfate (SDS) to improve stability. The solution was heated for 4 h to finish chelating reaction with Gd^{3+} and

Y. Liu, *Multifunctional Gold Nanostars for Cancer Theranostics*, Springer Theses,
https://doi.org/10.1007/978-3-319-74920-4_2

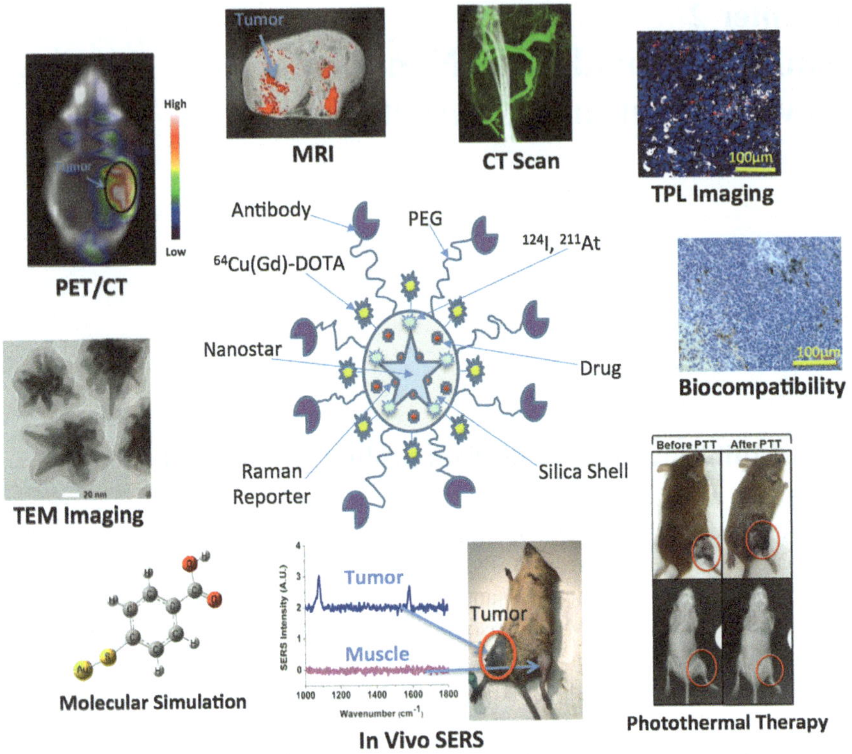

Fig. 2.1 Illustration of the multifunctional GNS nanoprobe for cancer imaging and nanotherapy

then nanoparticles were centrifuge washed three times to remove all non-chelated Gd^{3+}. The DOTA chelators could also be used to bind $^{64}Cu^{2+}$ positron emitter for PET imaging. In addition, we also tested ^{131}I and ^{211}At labeling on the GNS nanoprobe with PEG polymer coating. They are therapeutic radioisotopes used for cancer treatment. The reaction was performed by mixing radioisotopes with PEGylated GNS nanoparticles in 20 mM of NaOH solution with 30 min incubation at room temperature.

Synthesized GNS nanoprobe concentration, size distribution and ζ-potential were measured with Nanoparticle Tracking Analysis (NTA) method by using the NanoSight NS 500 (Amesbury, U.K.). Raman spectra measurement was performed on a HORIBA Jobin Yvon LabRAM ARAMIS system (Edison, NJ) with 785 nm excitation (40 mW) and 10 s integration time. Transmission electron microscopy (TEM) images were acquired on an FEI Tecnai G2 Twin transmission electron microscope (Hillsboro, OR) with 160 kV accelerating voltage. Multifunctional GNS nanoprobes in solution were tested using a 1.5 T whole body clinical MR scanner (GE Healthcare, WI) with a standard head coil. An inversion recovery sequence (*TE*/*TR* = 9/5000 ms) with inversion times of 500 and 2100 ms was used to calculate the T_1 relaxation times. CT images were acquired on a clinical CT scanner (GE

Healthcare, WI, USA) with submilimeter spatial resolution. PET/CT scan was performed by using a μPET/CT scanner (Siemens Medical Systems, TN, USA).

2.2 Multifunctional GNS Characterization

As shown in Fig. 2.2, the synthesized GNS multifunctional nanoprobe is composed of a 60 nm gold nanostar with a 20 nm silica shell and has a plasmonic extinction maximum at approximately 850 nm, which is in the NIR tissue optical window. The hydrodynamic diameter and ζ-potential were measured to be 73 nm and −5.7 mV for the PEGylated GNS nanoparticles and 110 nm and −42.2 mV for the silica coated GNS multifunctional nanoprobe with Gd^{3+} loading. The number of Gd^{3+} ions per GNS nanoprobe was determined to be 64,200 ± 320 (mean ± S.D.)

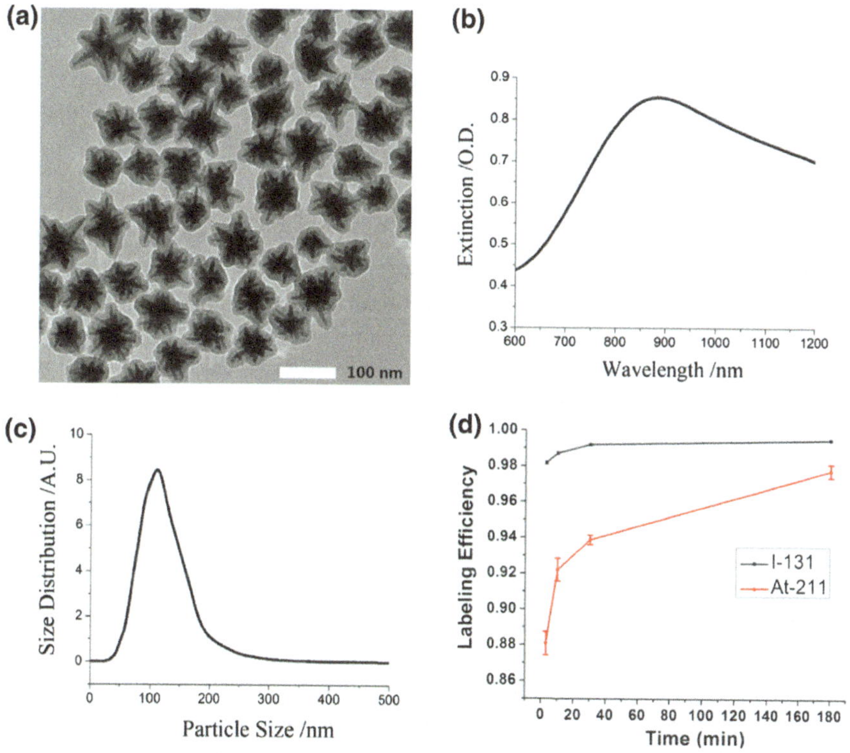

Fig. 2.2 **a** TEM imaging of the developed multifunctional GNS nanoprobes. **b** The extinction spectrum of 0.1 nM GNS nanoprobes in water solution. **c** Hydrodynamic size distribution of GNS nanoprobes. **d** Radiolabeling efficiency of GNS with ^{131}I or ^{211}At after 3, 10, 30 and 180 min incubation at room temperature. Three runs for each time point were performed and the error bar is shown as the standard deviation. (Adapted from Ref. [10])

by using inductively coupled plasma-mass spectroscopy (ICP-MS) on a Varian ICP-MS 820. The batch-to-batch variation was estimated to be 4% from three independent runs. Covalent-bound Gd^{3+} ions on the nanoprobe were examined by acquiring T_1-weighted magnetic resonance images of the supernatant after centrifugation washes. The calculated relaxivity of the GNS nanoprobe is 4.1×10^6 mM^{-1} s^{-1} in H_2O at a magnetic field strength of 1.5 T. The relaxivity value per Gd^{3+} ion is calculated to be 63.8 mM^{-1} s^{-1} and it is comparable with previous studies with values of 72.3 and 23 $mM^{-1}s^{-1}$ [3, 4]. The DOTA on the surface of GNS nanoprobe can also be used for other metallic contrast agent, such as $^{64}Cu^{2+}$ for PET imaging. Furthermore, the gold surface can be used to load iodine (I) or astatine (At) radioisotopes by I-Au or At–Au chemical bond [5]. After radiolabeling process, the GNS nanoprobes were centrifuge washed at 6000 rcf for 20 min three times and the radioactivity in the pellet and supernatant fluid was measured separately. The radiolabeling efficiency was calculated by the ratio of radioactivity in pellet to the total radioactivity in pellet and solution. ^{131}I and ^{211}At radioisotopes were labeled to GNS in 30 min with high yield of more than 90% (Fig. 2.2d). We also tested maximum capacity of GNS nanoprobe for ^{211}At therapeutic radioisotope loading. Three ^{211}At:GNS ratios were tested (140, 12 and 1.5). And radiolabeling efficiency after 30 min incubation at room temperature was calculated to be $86.6 \pm 0.6\%$, $95.5 \pm 0.2\%$ and $97.0 \pm 0.1\%$, respectively. It showed that the radiolabeling efficiency decreases slightly with increased ^{211}At:GNS ratio.

After nanoprobe development, we performed sensitivity evaluation for each modality. As shown in Fig. 2.3, the developed multifunctional GNS nanoprobe has a detection limit of 2 pM (SERS), 10 pM (MRI) and 100 pM (CT). The SERS method has higher detection sensitivity than CT and MRI. SERS takes advantage of localized surface plasmon and can enhance Raman signal up to 10^{13} to 10^{14} folds, allowing highly sensitive detection. The developed GNS nanoprobe has high Gd^{3+} loading capacity, resulting in high sensitivity under MRI scan down to 10 pM while typical 100 μmol kg^{-1} of Gd^{3+} contrast agent dose is administrated in clinical applications. CT imaging has relatively low contrast discrimination and 100 pM detection limit of GNS nanoprobe was measured. Gold contrast agent with higher atomic number can provide better contrast than iodinated contrast agent, which is currently used in medical imaging. However, even with contrast agent enhancement, CT imaging still has limits for clinical applications, especially for soft tissue imaging.

Furthermore, we also evaluated the sensitivity and spatial resolution of PET imaging (Fig. 2.4). PET can reach extremely high sensitivity, down to 10 fM of ^{124}I, which is several orders of magnitude higher than MRI or CT. It should be noted that for in vivo applications, PET imaging has almost zero background signal in normal tissues since there are no natural positron emitters in the human body. On the contrary, normal tissues have intrinsic signal under MRI or CT scan due to

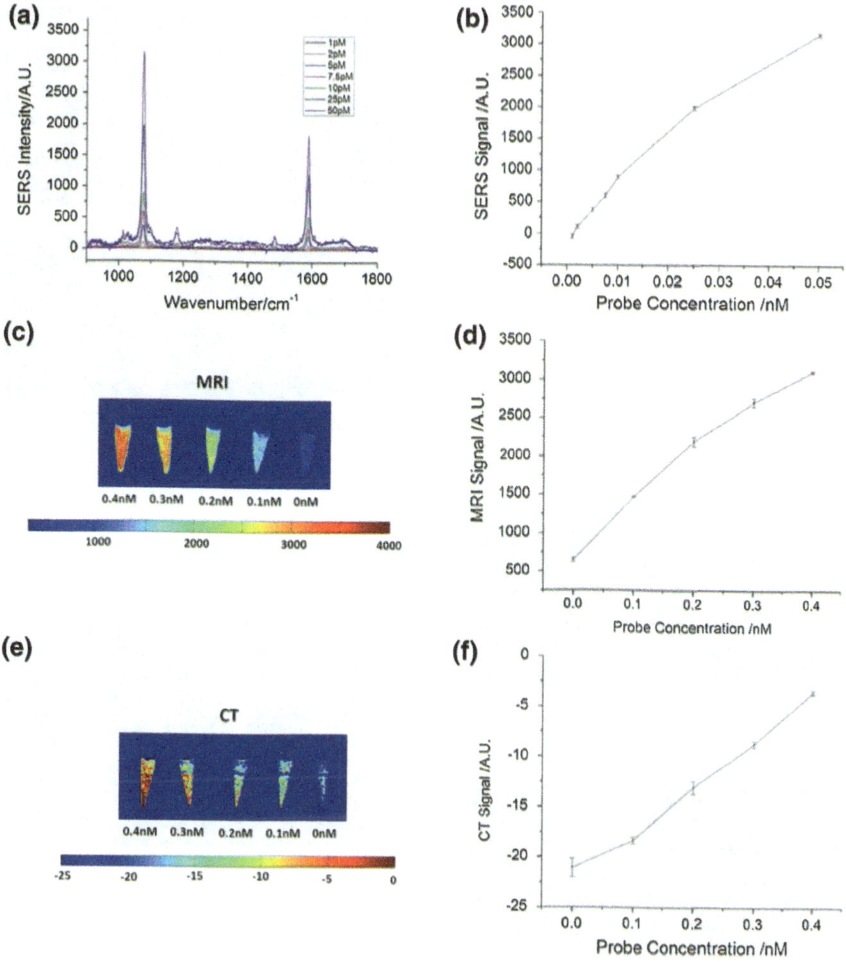

Fig. 2.3 **a** SERS spectra of GNS nanoprobes with different concentration. **b** SERS intensity at 1066 cm^{-1} measured in GNS solution with concentrations ranging from 1 pM to 25 pM. **c** MRI images of GNS nanoprobes with different concentration. **d** MRI T_1 intensity of nanoprobes in solution with concentrations ranging from 0 to 0.4 nM. **e** CT images of GNS nanoprobes with different concentration. **f** CT intensity of nanoprobes in solution with concentrations ranging from 0 to 0.4 nM. (Adapted from Re [10])

proton existence or X-ray attenuation. The spatial resolution of PET imaging was tested with 1 μl of ^{124}I labeled GNS in 1 ml agar gel. The radiolabeled GNS can be clearly identified and 1 mm spatial resolution can be reached. With low background signal, high sensitivity and spatial resolution, PET imaging provides a superior method for GNS nanoprobe in vivo tracking noninvasively in 3D.

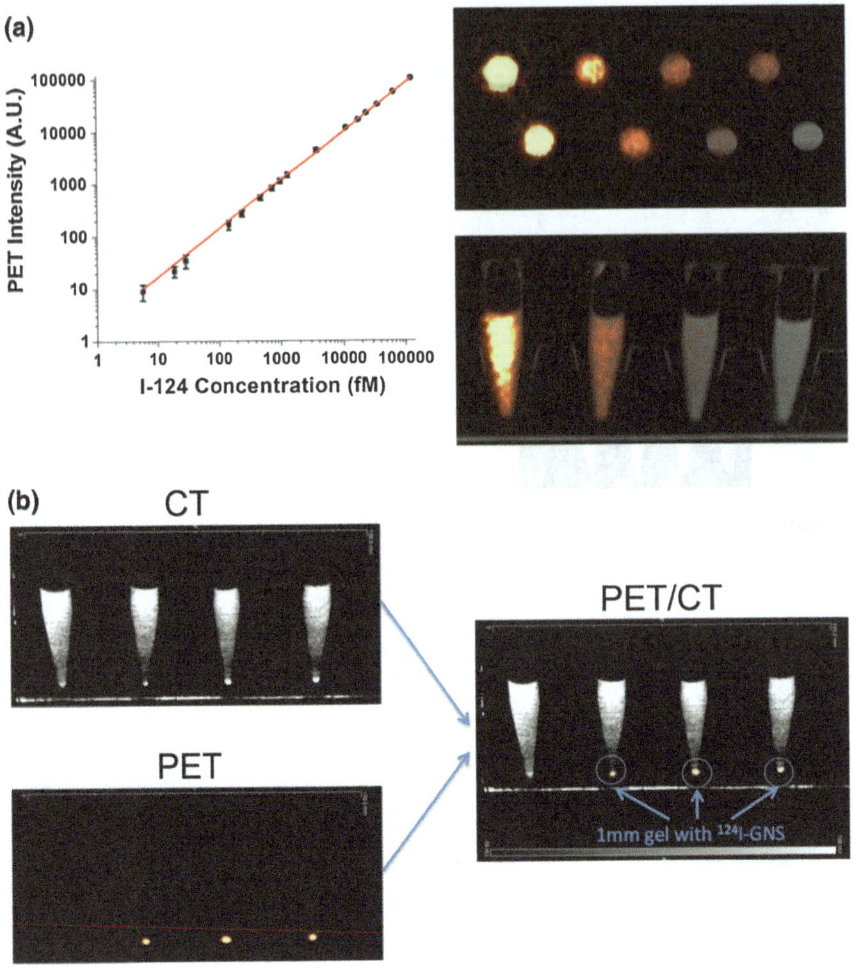

Fig. 2.4 PET sensitivity and spatial resolution evaluation with ^{124}I radioisotopes coated on GNS nanoprobe. **a** 1 ml samples containing different ^{124}I radioactivity was used for the sensitivity evaluation. 10 fM detection limit is achieved. **b** 1 μl gel sample with ^{124}I-GNS was prepared and then the tube was filled with 1 ml gel without ^{124}I-GNS. CT scan, PET scan and an overlap of PET and CT. The results demonstrate PET scan can detect a sample with 1 mm in size

2.3 pH Sensing with GNS Nanoprobe Using SERS

The pH sensing capability of GNS nanoprobe with pMBA reporter and PEG coating was tested with phosphate buffer at different pH values (Fig. 2.5). In addition, density functional theory (DFT) calculation was also performed to

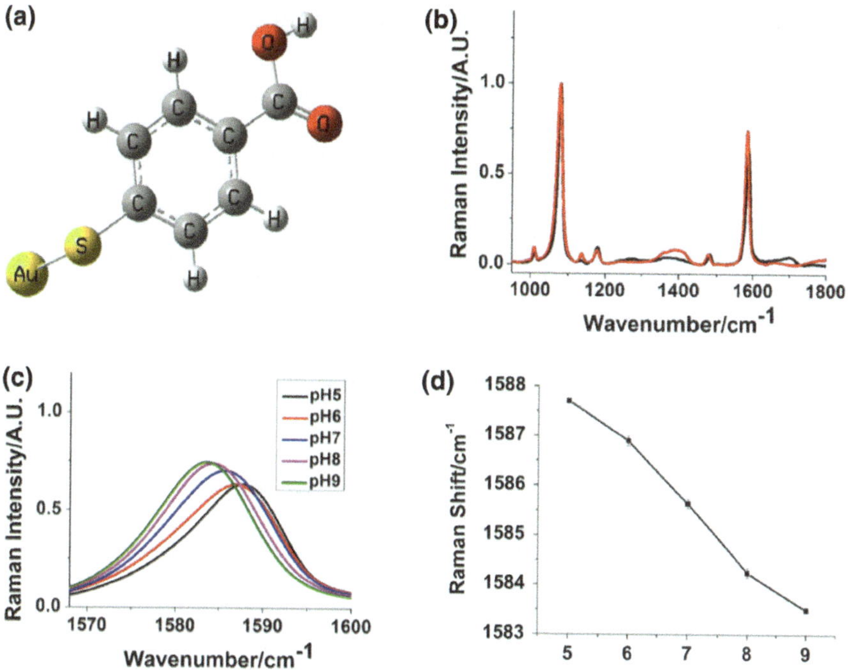

Fig. 2.5 **a** pMBA-Au complex structure used for density functional theory (DFT) calculation. **b** SERS spectra comparison obtained from GNS nanoprobes at pH 5 (Black) and 9 (Red). The spectrum is normalized with intensity at 1078 cm^{-1}. **c**, **d** SERS peak position shift between pH 5 and 9. (Adapted from Ref. [6])

investigate the vibrational mode of SERS peaks of interest. The acquired SERS spectrum at pH 5 and pH 9 is slightly different, especially for the normalized intensity of the peak at 1390 and 1700 cm^{-1}. The peak intensity at 1390 cm^{-1} increase as the pH changes from 5 to 9. On the contrary, the peak intensity at 1700 cm^{-1} decreases when the pH changes from 5 to 9. The Raman peak at 1390 and 1700 cm^{-1} was identified to be due to deprotonated and protonated carbolic group stretching vibrational mode, respectively [6, 7]. Therefore, the intensity of those two peaks has different change trends with pH. Furthermore, a novel pH sensing index, the SERS peak shift at ~1580 cm^{-1} was observed experimentally and explained with theoretical calculations. The peak position shift was found to be due to the vibrational mode coupling between the benzene ring and carboxylic group stretching. The GNS multifunctional nanoprobe has potential to be used for pH sensing in future biomedical applications.

2.4 GNS Photothermal Conversion Evaluation

The photothermal conversion of GNS nanoprobe was evaluated for future in vivo photothermal therapy to treat cancer. GNS transducers have no decrease in the photon-to-heat conversion efficiency after three repetitive 10 min laser irradiations with power density of 0.8 W/cm^2 (Fig. 2.6a). Figure 2.6b shows that equilibrium temperature varies with GNS concentration with the same laser power density. Water alone shows minimal light absorption at 808 nm laser wavelength with 3.1 °C temperature increase while 21.7 °C increase for 50 µg/ml GNS solution. Given the same mass, the photon-to-heat conversion capability of 30 nm GNS is 96,000 times higher than that of water. It demonstrates GNS nanoprobe is a superior photon-to-heat transducer. Furthermore, a comparison experiment was performed

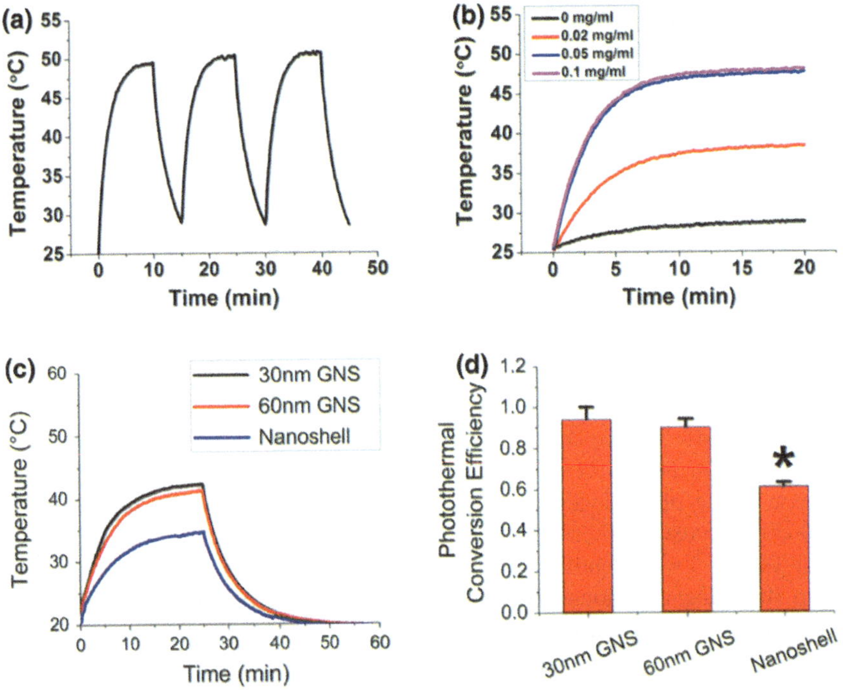

Fig. 2.6 Photothermal evaluation of GNS in solution. **a** Repetitive photothermal heating (50 µg/ml 30 nm GNS nanoprobe in water solution, 0.8 ml total solution volume) with 0.8 W/cm^2 of 808 nm laser turning on at 0, 15, and 30 min and laser turning off at 10, 25, and 40 min. **b** Temperature profile for different GNS concentrations with the same 0.8 W/cm^2 laser power density. **c** Temperature profile of 30 nm GNS, 60 nm GNS, and gold nanoshell for photothermal conversion efficiency calculations. **d** Calculated photothermal conversion efficiency for GNS nanoprobe and gold nanoshell as a comparison (n = 3 for each group). Both 30 and 60 nm GNS have higher photothermal conversion efficiency than gold nanoshell nanoparticles. (Adapted from Ref. [2])

to evaluate photothermal conversion efficiency for 30 nm GNS, 60 nm GNS and gold nanoshell, which is one of the most well studied nanoparticles and currently under clinical trials for photothermal therapy. Figure 2.6c shows the temperature profiles for each of these three nanoparticles, with the 30 nm and 60 nm GNS having a much higher equilibrium temperature (42.3 and 41.4 °C) than nanoshells (34.7 °C) at equivalent optical density. The calculated values of photothermal conversion efficiency (Fig. 2.6d) were 94% for 30 nm GNS, 90% for 60 nm GNS, and 61% for gold nanoshell. The measured conversion efficiency of gold nanoshell is consistent with previously published results [8, 9]. The difference in efficiencies between the gold nanoshell and the two sizes of GNS was found to be statistically significant. Experiment results demonstrate that GNS have a higher efficiency in converting the delivered laser light into heat, which is superior for photothermal treatment.

2.5 Summary

We have developed a multifunctional GNS nanoprobe for SERS, MRI, CT, PET, PTT and therapeutic radioisotope delivery. The GNS nanoprobe has high SERS enhancement and could be used for in vivo sensing. PET imaging with GNS nanoprobe has the highest sensitivity and could be used for sensitive tracking. In addition, the superior ^{211}At loading and photon-to-heat conversion efficiency show GNS nanoprobe's potential for cancer treatment. As a result, the GNS multifunctional nanoprobe, with capabilities for both cancer imaging and therapy, has potential to be used for future cancer management such as early cancer detection and image-guided specific therapy.

References

1. Yuan H, Khoury CG, Hwang H, Wilson CM, Grant GA, Vo-Dinh T (2012) Gold nanostars: surfactant-free synthesis, 3D modelling, and two-photon photoluminescence imaging. Nanotechnology 23(075102):1–7
2. Liu Y et al (2015) A plasmonic gold nanostar theranostic probe for in vivo tumor imaging and photothermal therapy. Theranostics 5:946–960
3. Gerion D et al (2007) Paramagnetic silica-coated nanocrystals as an advanced MRI contrast agent. J Phys Chem C 111:12542–12551
4. Courant T et al (2012) Hydrogels incorporating GdDOTA: towards highly efficient dual T1/T2 MRI contrast agents. Angewandte Chemie-Int Ed 51:9119–9122
5. Schulz A, Hargittai M (2001) Structural variations and bonding in gold halides: A quantum chemical study of monomeric and dimeric gold monohalide and gold trihalide molecules, AuX, Au_2X_2, AuX_3, and Au_2X_6 (X = F, Cl, Br, I). Chem-A Eur J 7:3657–3670
6. Liu Y, Yuan H, Fales AM, Vo-Dinh T (2013) pH-sensing nanostar probe using surface-enhanced Raman scattering (SERS): theoretical and experimental studies. J Raman Spectrosc 44:980–986

7. Michota A, Bukowska J (2003) Surface-enhanced Raman scattering (SERS) of 4-mercaptobenzoic acid on silver and gold substrates. J Raman Spectrosc 34:21–25
8. Ayala-Orozco C et al (2014) Au nanomatryoshkas as efficient near-infrared photothermal transducers for cancer treatment: benchmarking against nanoshells. ACS Nano 8:6372–6381
9. Cole JR, Mirin NA, Knight MW, Goodrich GP, Halas NJ (2009) Photothermal efficiencies of nanoshells and nanorods for clinical therapeutic applications. J Phys Chem C 113:12090–12094
10. Liu Y, Chang Z, Yuan H, Fales AM, Vo-Dinh T (2013) Quintuple-modality (SERS-MRI-CT-TPL-PTT) plasmonic nanoprobe for theranostics. Nanoscale 5:12126–12131

Chapter 3
In Vivo Evaluation of GNS Nanoprobe

3.1 Background

[1]GNS nanoparticles' in vivo biodistribution, tumor uptake and pharmacokinetics are of great importance for their biomedical applications [3–6]. Nanoparticles take advantage of EPR effect from tumor leaky vasculature for cancer targeting. Particle size plays an important role in determining the biodistribution and clearance pathway (via liver/spleen or kidney). Nanoparticles with size less than 10 nm have been found to be cleared by the kidney, while those greater than 200 nm are quickly filtered by the liver and spleen [7, 8]. Consequently, a nanoparticle size between 30 and 200 nm has been considered to be ideal for accumulation in tumors through the EPR effect. Surface charge can also affect nanoparticles' in vivo behavior. Nanoparticles with negative or positive surface charge could absorb plasma protein and get cleared quickly after IV injection. On the contrary, nanoparticles with polyethylene glycol (PEG) coating have neutral surface and been reported to lengthen circulation time, which is preferable for tumor accumulation with EPR effect [9–12]. PET, MRI and CT imaging modalities have been used for tracking nanoparticles at the whole body level while optical imaging, with higher spatial resolution, can be used to track nanoparticles at the subcellular level [13–19]. The GNS multifunctional nanoprobe, with its intrinsic high TPL signal and capabilities for PET, MRI and CT imaging, provides a powerful tool to investigate how the nanoparticles' properties affect their in vivo behaviors.

[1]Parts of this chapter were adapted from the References [1, 2].

Y. Liu, *Multifunctional Gold Nanostars for Cancer Theranostics*, Springer Theses,
https://doi.org/10.1007/978-3-319-74920-4_3

3.2 PET Imaging with Radiolabeled GNS Nanoprobe

First, we performed dynamic PET/CT scan with radiolabeled GNS nanoprobe to show its capability as a PET nanoprobe. The GNS nanoprobes were labeled with $^{64}Cu^{2+}$ radioisotopes by DOTA chelators linked to the silica surface of GNS nanoprobe. The labeling efficiency was calculated to be 86.8 ± 6.6% ($n = 3$). The stability of the $^{64}Cu^{2+}$ DOTA chelation was studied in mouse plasma before in vivo application. Following 24 h of incubation at 37 °C, greater than 95% of the radioactivity was retained on the nanoprobes. As shown in the Fig. 3.1a, 2-h continuous PET scan results revealed an immediate nanoprobe uptake in the liver, which is the major organ of the reticuloendothelial system (RES). The nanoprobe concentration in blood increased quickly to 34.6%ID/g at 0.5 min after IV injection. At the same time, liver uptake increased to 4.7%ID/g. At 2 min, the liver

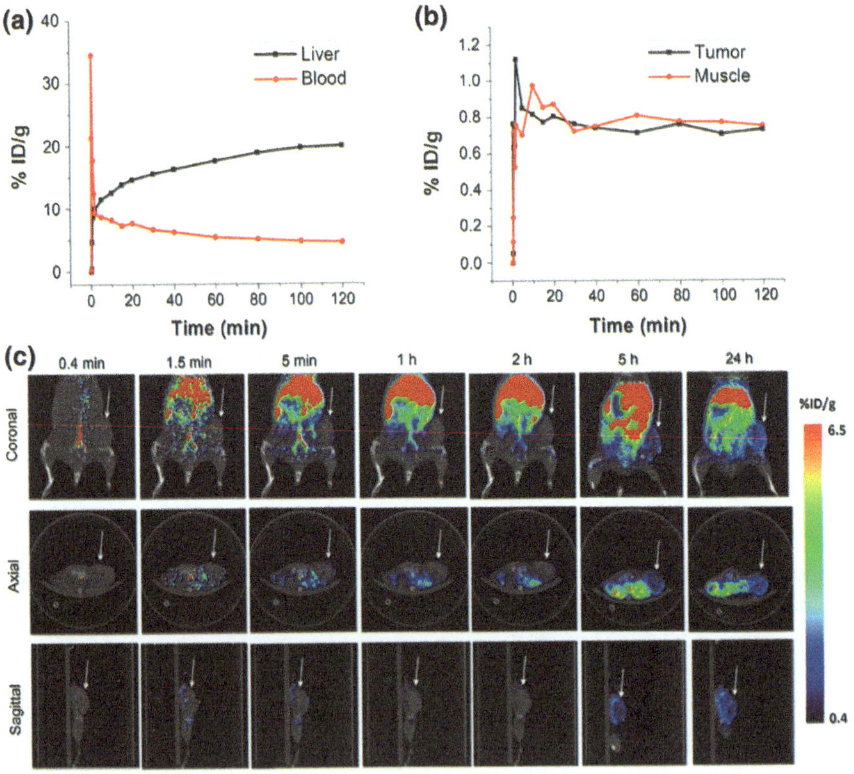

Fig. 3.1 Measured radioactivity in liver, blood (**a**) and tumor, muscle (**b**) after IV administration of ^{64}Cu labeled GNS nanoprobe. **c** Imaging of PET/CT on a mouse with flank tumor after systemic nanoprobe injection. Images were obtained at different times points. White arrow shows tumor location. The GNS uptake in the tumor periphery increased gradually within 24 h after IV injection. Adapted from reference [19]

uptake increased rapidly to 9.85%ID/g while the nanoprobe in blood decreased dramatically to 12.5%ID/g. The rapid decrease of nanoprobe concentration in blood could partially be due to the filtration effect of RES organs including liver. In the initial 2 h, tumor and muscle uptake were quite similar as shown in Fig. 3.1b. The nanoprobe uptake increased quickly to 0.85%ID/g (tumor) and 0.75%ID/g (muscle) at the end of 2 min. Then, the nanoprobe uptake in tumor and muscle fluctuated at around 0.8%ID/g until the end of 2 h. Between 2 and 24 h, the tumor uptake of nanoprobe increases slowly with time. Although no preferential accumulation in tumor was observed in the initial 2 h time period, the tumor-to-muscle ratio increased to 2.5:1 and 3.3:1, at the end of 5 and 24 h, respectively. The results demonstrate radiolabeled GNS nanoprobe can be used as a PET contrast agent for in vivo imaging.

PEG polymer coating has been reported to be able to improve nanoparticles' in vivo stability and increase blood circulation time to enhance tumor uptake with EPR effect. Therefore, we also tested the PEGylated GNS for PET imaging. 30 nm GNS nanoprobes with PEG coating were labeled with ^{124}I radioisotopes by Au-I chemical bond. The labeling efficiency was greater than 98% after 30 min incubation at room temperature. The stability of radiolabeled GNS was examined in both phosphate buffered saline (PBS) and plasma with anti-clotting agent (heparin), and experimental results showed that 97.2 ± 0.2% (PBS) and 97.7 ± 0.4% (plasma) of ^{124}I was found to remain on GNS 7 days after 37 °C incubation. Figure 3.2 illustrates in vivo pharmacokinetics of PEGylated 30 nm GNS nanoprobe analyzed with the two-compartment model by fitting relative plasma concentration measured in the

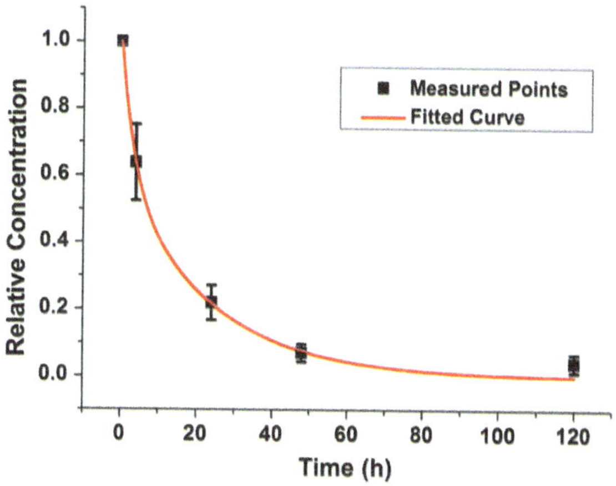

Fig. 3.2 GNS pharmacokinetics study with PET scan. 30 nm PEGylated GNS nanoprobe was labeled with ^{124}I for PET imaging. Nanoprobe relative concentration in blood measured from dynamic PET/CT scans at 10 min, 4, 24, 48 and 120 h after intravenous injection through the mouse-tail vein (normalized by the concentration at 10 min). The results were fitted with the two-compartment pharmacokinetic model. The adjusted R^2 for the fitted curve is 0.991. The terminal half-life ($T_{1/2}$) was calculated to be 16.3 h. 3 mice were used in this study and the error bar is shown as the standard deviation

heart from PET/CT scan at five time points (10 min, 4, 24, 48 and 120 h). The terminal half-life ($T_{1/2}$) of I-124 labeled GNS serum clearance was calculated to be 16.3 h. Error bar is shown as standard deviation and the adjusted R^2 for the fitting is 0.991. The measured circulation half-life is similar to 15 × 50 nm PEGylated nanorod (17.6 h), 110 nm PEGylated liposomes (16.5 h) and 20 nm micelles (15.5 h) [20, 21]. PEG coated GNS showed improved pharmacokinetic property over silica coated GNS and is preferable for in vivo imaging and drug delivery.

3.3 CT Imaging with GNS Nanoprobe

The GNS multifunctional nanoprobe can also be used as a contrast agent for CT imaging. CT is one of the most commonly used imaging modalities in the clinic and is also useful for tumor detection and treatment planning for both surgery and radiation therapy. CT imaging also provides non-invasive real time dynamic information and allows us to visualize which regions of the tumor are well vascularized and which regions experience little nanoparticle delivery. This information can therefore be very useful to improve our understanding of nanoparticles' in vivo performance for efficient drug delivery. PEGylated GNS nanoprobes with different sizes (30 and 60 nm) were used with dual energy CT imaging on mice with primary sarcomas to investigate in vivo dynamic biodistribution and demonstrate real-time monitoring of intratumoral accumulation. A custom-built dual-energy micro-CT imaging system was used for this study. Two acquired datasets (80 and 40 kVp energies) were reconstructed using the Feldkamp algorithm, co-registered, and smoothed using joint bilateral filtration. Dual energy decomposition was performed by fitting the voxel data to a CT sensitivity matrix derived from gold concentration standards, which we have previously validated in vivo. The result of the dual energy decomposition was a gold concentration map, in which intensity values represent the calculated concentration of gold within each voxel. Following segmentation, gold concentrations were measured in each region of interest by calculating the average value of the gold maps over the entire region of interest.

The longitudinal dual-energy CT imaging, 30 min, 24, and 72 h after gold nanoparticles IV injection, showed gold nanoparticle distribution at various time points (Fig. 3.3) [1]. Due to low contrast discrimination of CT, a high GNS dose (800 mg/kg) was injected to get imaging up to a higher quality. The CT images at 30 min clearly show gold nanoparticles localized in blood vessels. It appears that the majority of blood vessels in the tumor region were located on the outside of the tumor, so perfusion to the core of the tumor may be limited. At the 24 and 72 h time points, there was nanoparticle accumulation in the tumors. CT results also show a comparison of the concentrations of gold within the tumor rim and tumor core. CT slices through the tumors show this heterogeneous intratumoral distribution (Fig. 3.4). The gold concentration at the rim is much higher than that in the center

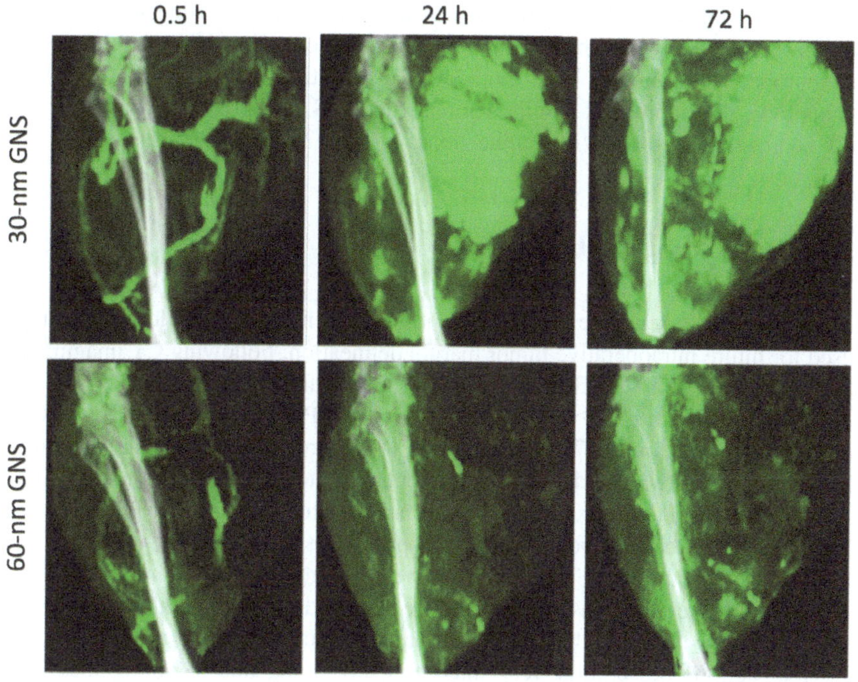

Fig. 3.3 Maximum intensity projection of CT images for primary sarcoma after GNS IV injection. Gold is shown in green color ranging from 2 to 10 mg/ml. Other tissues are shown in gray scale ranging from -500 to 1200 HU. Adapted from reference [1]

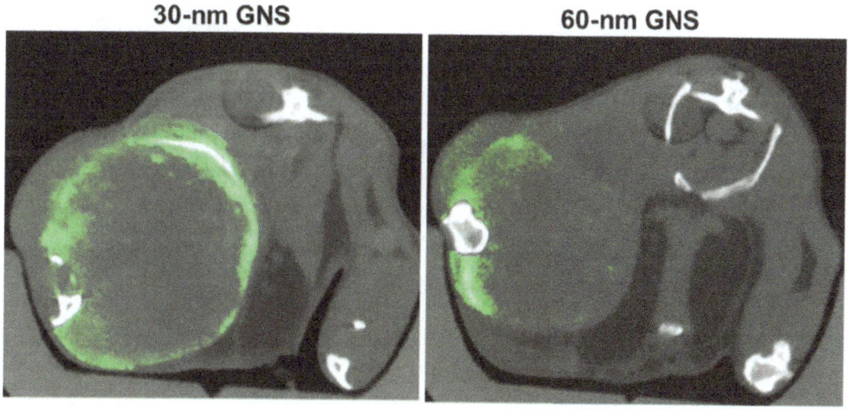

Fig. 3.4 Dual-energy CT scan on mice with primary sarcomas 72 h after IV administration of 30 nm or 60 nm GNS. The axial slices are shown. Gold is shown in green ranging from 2.5 to 8.0 mg/ml and all other tissues are shown in gray scale ranging from −500 to 1200 HU. Adapted from reference [1]

and is not evenly distributed around the tumor rim. The low concentration at the center may reflect poor perfusion in the tumor core. In addition, 30 nm GNS nanoprobes are found to have more even distribution than that of 60 nm GNS nanoprobes. Experimental results demonstrate GNS nanoprobes can be used as a CT contrast agent for in vivo applications.

3.4 In Vivo SERS Detection

SERS provides a sensitive method for molecular sensing and could be potentially used for future intraoperative tumor margin delineation. Following CT imaging, the SERS spectra from the 30 nm GNS that accumulated within each tumor and contralateral leg muscle were measured with a PIXIS: 100BR_eXcelon CCD mounted to an Acton LS-785 spectrograph (1200 grooves/mm grating), controlled by LightField software (Princeton Instruments, Trenton, NJ). A 785 nm diode laser was fiber-coupled to the excitation fiber of an InPhotonics Raman probe (Norwood, MA) and the collection fiber of the Raman probe was coupled to the entrance slit of the LS-785 spectrograph. SERS spectra were recorded for mice in the CT study immediately after euthanasia. Spectra were recorded for all mice through the skin over the tumor and over the leg muscle on the contralateral side as a control.

Gold nanoprobes exhibit very intense SERS signal due to strong local field enhancement at the tips of the nanostar spikes. Our previous study showed that GNS SERS enhancement is more than two orders of magnitude higher than that of gold nanospheres. Figure 3.5 shows the SERS spectrum measured in the sarcoma

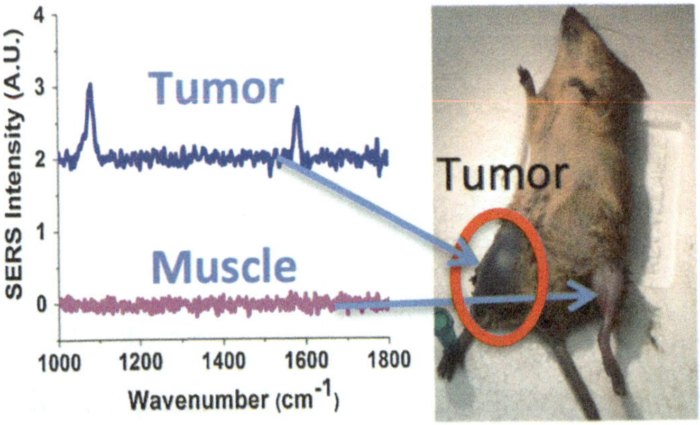

Fig. 3.5 SERS spectra of 30 nm GNS nanoprobe with pMBA Raman reporter measured from mice with primary sarcoma. The measurement was performed 3 days after IV injection of 30 nm GNS nanoprobe. The SERS peaks of pMBA reporter can only be observed in the tumor, but not in the normal muscle of the contralateral leg. Significant amount of GNS nanoprobe accumulated in the tumor, but not in the contralateral normal leg. The tumor is shown in black color due to high GNS uptake. Adapted from reference [1]

and normal muscle [1]. Tumor part was shown as black color due to a high amount of GNS accumulation. The GNS were labeled with pMBA, a Raman reporter. The characteristic SERS peaks of pMBA on GNS at 1067 and 1588 cm^{-1} were detected in the tumor, but not in the contralateral leg muscle, which shows that SERS has the capability to differentiate tumor from normal muscle. Those two peaks with high intensity were assigned to be the stretching vibrational mode of the benzene ring in pMBA based on previous combined theoretical and experimental investigations. The in vivo SERS experiment results demonstrate our GNS nanoprobe has potential to be used for tumor differentiation for clinical applications.

3.5 TPL for GNS Tracking at Subcellular Level

GNS nanoprobes with tip-enhanced plasmonics, have extremely strong two-photon photoluminescence and can be used to monitor nanoparticle distribution at the cellular level without dye labeling. The calculated two-photon absorption cross-section for branched GNS is approximately 5×10^4 times higher than for equivalent-sized gold nanospheres, which means that even single GNS appears very bright when imaged with two-photon microscopy [22]. After CT imaging, tumors were sectioned, immunostained and imaged using a two-photon microscope. TPL imaging on immunostained tumor sections was performed using an Olympus FV 1000 multiphoton microscope (Olympus America, Center Valley, PA). The pulsed laser used was a Ti:Sapphire laser (Chameleon Vision II; Coherent, Santa Clara, CA) with tunable range of 680–1080 nm, 140-fsec pulse width, and 80 MHz repetition rate. The laser beam (800 nm wavelength) was focused with a water-immersion objective with 25X 1.05 NA (Olympus America, Center Valley, PA). Images were acquired for each channel (red, green, and blue) and were merged for the final TPL images. As shown in Fig. 3.6, both the 30 nm and 60 nm GNS nanoprobes exhibit extremely strong two-photon photoluminescence [1]. The emitted light after absorbing two photons in the NIR region, spans the visible spectrum, so the nanoparticles appear bright white. It can be seen that the 30 nm GNS nanoprobes penetrate more deeply into the tissue than 60 nm GNS after leaking through tumor vasculature. In addition, the 30 nm GNS have a fairly even distribution in tumor tissue, while the larger 60 nm GNS were restricted primarily to the perivascular space. The experimental results show GNS nanoprobes are superior for tracking with both high sensitivity and spatial resolution using two-photon microscopy.

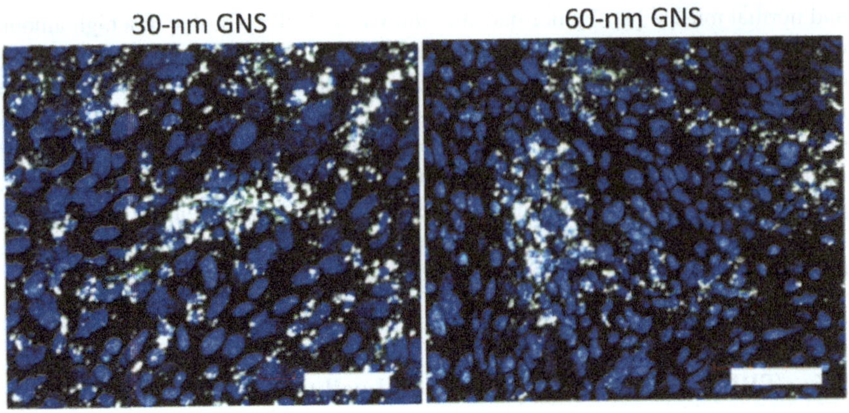

Fig. 3.6 High spatial resolution TPL imaging for GNS nanoprobe tracking at subcellular level. Both 30 and 60 nm GNS are identified in tumor. 30 nm GNS penetrate deeper into tumor interstitial space than 60 nm GNS after leaking through tumor vasculature. Scale bar, 50 μm. Blue, DAPI stained cell nuclei; green, dye stained CD31 showing vasculature; bright white, GNS nanoprobes. Adapted from reference [1]

3.6 GNS Biodistribution Study with Radiolabeling

Nanoparticles' biodistribution is of significant importance for their applications including drug delivery. In this study, we investigated how particle size and injection dose affects GNS nanoprobes' biodistribution. Furthermore, histopathology examination was also performed to show how GNS nanoprobes were cleared from blood circulation after IV injection through tail vein. PEGylated GNS (30 or 60 nm) were incubated with ^{131}I (5 μCi per 200 μg gold) in phosphate buffered saline (PBS) for 2 h at room temperature. Unbound ^{131}I was removed by pelleting and resuspending the nanoparticles in sterile PBS. The retention of ^{131}I on the GNS surface was investigated in PBS with 4% albumin. After 96 h incubation at 37 °C, the solution was centrifuged and the radioactivity of the supernatant and pellet was measured to determine the percentage of ^{131}I remaining on the GNS surface. Freshly radiolabeled GNS were used for biodistribution studies. Twelve mice with xenograft sarcomas were randomly divided into 3 groups (n = 4 in each group). Nanoparticles in PBS were injected intravenously via the tail vein. For the first group, 100 μl of radiolabeled 30 nm GNS (50 μg gold) was injected. For the second group, 100 μl of radiolabeled 30 nm GNS (200 μg gold) was injected. For the third group, 100 μl radiolabeled 60 nm GNS (200 μg gold) was used. The injection standard (5%) for each dose was prepared by diluting 100 μl of radiolabeled GNS solution to 2 ml with water. Mice were sacrificed 48 h after injection and organs of interest were harvested for radioactivity measurement. The radioactivity of each organ and internal standard was measured with an automated gamma counter (LKB 1282, Wallac, Finland), and the measured results were used to calculate percentage of injected dose per gram tissue (% ID/g) for each organ of interest.

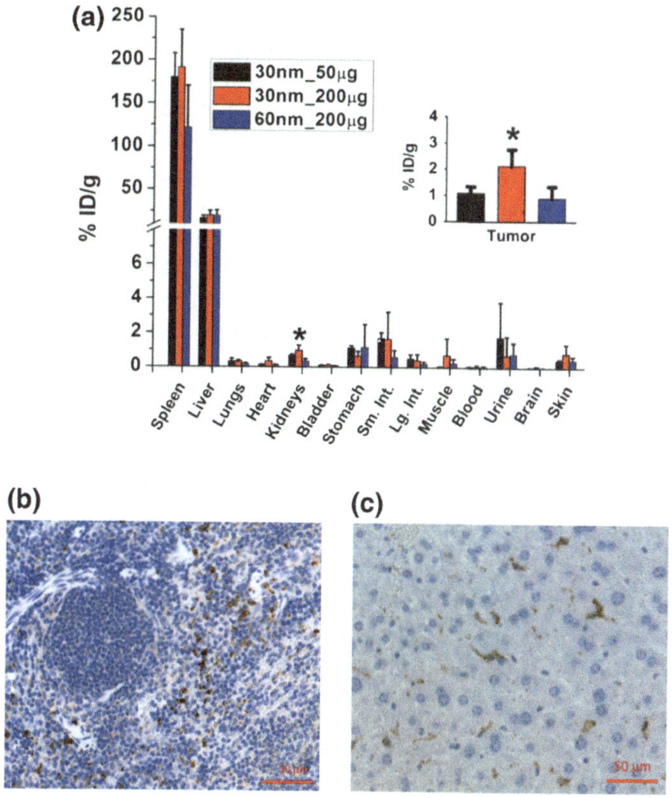

Fig. 3.7 **a** In vivo biodistribution of GNS nanoprobes 48 h after IV injection through tail vein. 30 nm GNS and 60 nm GNS were investigated in this study. In addition, two doses (50 and 200 μg) were chosen for 30 nm GNS for comparison. The asterisk represents a statistically significant difference from the other two groups ($p < 0.05$). F4/80 immunohistochemistry (IHC) staining (brown color) was performed for spleen (**b**) and liver (**c**) to show macrophage cells. Most of GNS nanoparticles (black spots) are found to be overlapped with macrophages cells (orange color). Adapted from reference [1]

As shown in Fig. 3.7a, tumor uptake of 30 nm GNS was higher (2.11 ± 0.64% ID/g) than that of 60-nm GNS (0.88 ± 0.46%ID/g) with the same injection dose of gold (200 μg) [1]. In addition, 30 nm GNS showed a higher tumor uptake with a 200-μg injection dose (2.11 ± 0.64%ID/g) than with 50-μg injection dose (1.08 ± 0.25%ID/g). This difference was found to be statistically significant. The reticuloendothelial system (RES), including liver and spleen, demonstrated high uptake of nanoparticles, which is consistent with previous studies [7, 23]. Histopathology examination was performed on liver and spleen and results show that GNS nanoparticles were cleared by macrophage cells (Fig. 3.7b, c). Relatively low uptake levels of nanoparticles were seen in other non-target organs. Especially for brain, minimal GNS uptake was found due to the intact blood-brain barrier

(BBB). In this study, we found the tumor uptake of the GNS to be 1–2%ID/g. A recent in vivo study on ^{198}Au-doped gold nanoparticles with similar PEG coating reported that nanospheres and nanodisks have much higher tumor uptake than that of nanorods and nanocages one day after IV injection [23]. The tumor uptake of different shaped nanoparticles was reported to be 23.2%ID/g (nanospheres), 2.0% ID/g (nanorods), 7.5%ID/g (nanocages) and 4.9%ID/g (nanodisks) at the end of 24 h. The differences seen in tumor uptake can be attributed not only to differences in shape, but also to differences in tumor type and injection dose. The previous study used EMT6 murine mammary carcinoma cells, while we used a xenograft sarcoma tumor model. The liver uptake for different nanoparticles was reported to be 34.9%ID/g (nanospheres), 55.0%ID/g (nanodisks), 52.1%ID/g (nanorods), and 63.4%ID/g (nanocages) at the end of 24 h [23]. The liver uptake of GNS is relatively low: 16.54 ± 2.97%ID/g (30 nm GNS, 50 μg), 19.77 ± 5.76%ID/g (30 nm GNS, 200 μg) and 19.49 ± 6.36%ID/g (60 nm GNS, 200 μg) at the end of 48 h. The lower liver uptake in this study might be due to dense PEGylation, which improves the ability of the GNS to evade phagocytosis. Therefore, GNS nanoprobes with smaller nanoparticle size, dense PEG coating and higher injection show better tumor accumulation.

3.7 In Vivo Photothermal Therapy

In vivo photothermal therapy was performed on mice bearing primary sarcomas to evaluate GNS nanoprobes' potential for cancer treatment. A 980-nm continuous-wavelength laser was used for the in vivo photothermal therapy test. The tumor temperature during laser irradiation was continuously and remotely monitored with an infrared thermal imaging camera (Thermo Tracer TS 7302, NEC, Japan). In addition, a thermocouple temperature probe was also used to measure tumor temperature before and after laser irradiation to verify temperature measured from an infrared thermal imaging camera. Two groups of mice with primary sarcomas (n = 2 in each group) were used. For the treatment group, 2 mg of 30 nm GNS were injected two days before photothermal therapy, while PBS solution was injected for the control group. The tumors were irradiated at 0.7 W/cm^2 for 10 min on both the front and back sides. Afterwards, the mice were monitored and the photothermal therapy effect was evaluated by observing the tumor size change.

As shown in Fig. 3.8a, NIR images depicted tumor temperature change during the photothermal treatment process and showed temperature increase of sarcomas with and without GNS over 10 min of laser irradiation [1]. The tumor temperature is much higher for mice with GNS injection than for mice with PBS injection (control). The tumor surface temperature with GNS reaches 50 °C after only 4 min of treatment, which is high enough to kill tumor cells (Fig. 3.8b). The tumor treated with GNS and laser irradiation regressed to an undetectable size one day after

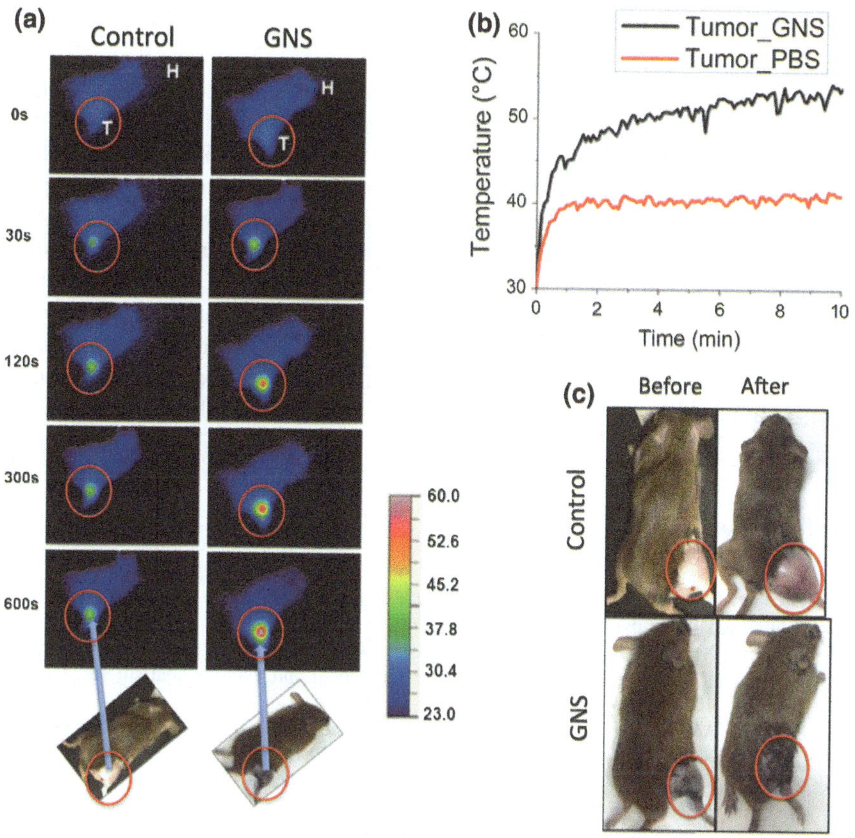

Fig. 3.8 **a** Tumor temperature monitoring with near-infrared camera for the mouse with sarcoma under laser irradiation after intravenous injection of 30 nm GNS or PBS solution (control). T shows the mouse tumor location and H represents the mouse head. **b** Tumor temperature change profile for the mouse under laser irradiation with IV injection of 30 nm GNS or PBS (control). It can be seen that the tumor temperature is significantly higher with GNS administration than control. **c** Photographs of mice before and after photothermal therapy. It shows a clear decrease in tumor size for the mouse with GNS injection. On the contrary, the tumor size in control mouse increased dramatically. Adapted from reference [1]

photothermal therapy with GNS, while tumors treated with PBS and laser irradiation continued to grow rapidly. As shown in Fig. 3.8c, the mice with GNS IV injection exhibited no sign of tumor growth for 7 days following treatment, at which point they were sacrificed. The laser irradiation was confined primarily to the tumor and there was no detectable tissue damage outside of the tumor region. There was some skin burning directly over the tumor surface, but no other adverse effects were observed in these mice.

3.8 Summary

The in vivo evaluation experiment results demonstrate the developed multifunctional GNS nanoprobe can be used for imaging with PET, CT, SERS and TPL as well as photothermal therapy for cancer treatment. Biodistribution study shows dense PEG coating, small size and high injection dose results in better tumor uptake of GNS nanoprobes. Pharmacokinetics study also shows GNS nanoprobes with a smaller size and dense PEG coating have sufficient in vivo stability and enlonged circulation time for tumor accumulation with EPR effect. In addition, high-resolution TPL imaging demonstrates GNS nanoprobes with smaller size can penetrate deeper into tumor interstitial space after leaking through tumor vasculature. Therefore, 30 nm GNS nanoprobes with dense PEG coating are chosen for further studies including sensitive brain tumor detection and photoimmunotherapy.

References

1. Liu Y et al (2015) A plasmonic gold nanostar theranostic probe for in vivo tumor imaging and photothermal therapy. Theranostics 5:946–960
2. Liu Y, Yuan H, Fales AM, Vo-Dinh T (2013) pH-sensing nanostar probe using surface-enhanced Raman scattering (SERS): theoretical and experimental studies. J Raman Spectrosc 44:980–986
3. Pekkanen AM, DeWitt MR, Rylander MN (2014) Nanoparticle enhanced optical imaging and phototherapy of cancer. J Biomed Nanotechnol 10:1677–1712
4. Chen C-C, Wu C-J, Yeh M-K (2013) Gold nanoparticle-based platforms as cancer-targeted molecules delivery systems. Int J Nanotechnol 10:840–849
5. Xie J, Lee S, Chen X (2010) Nanoparticle-based theranostic agents. Adv Drug Deliv Rev 62:1064–1079
6. Jain RK, Stylianopoulos T (2010) Delivering nanomedicine to solid tumors. Nat Rev Clin Oncol 7:653–664
7. Perez-Campana C et al (2013) Biodistribution of different sized nanoparticles assessed by positron emission tomography: a general strategy for direct activation of metal oxide particles. ACS Nano 7:3498–3505
8. Sykes EA, Chen J, Zheng G, Chan WCW (2014) Investigating the impact of nanoparticle size on active and passive tumor targeting efficiency. ACS Nano 8:5696–5706
9. Kim Y-H et al (2011) Tumor targeting and imaging using cyclic RGD-PEGylated gold nanoparticle probes with directly conjugated iodine-125. Small 7:2052–2060
10. Daou TJ, Li L, Reiss P, Josserand V, Texier-Nogues I (2008) Effect of ligand nature on the kinetics and the bio-distribution of quantum dots in mice. Nanotechnol 2008: Life Sci Med Bio Mater 2:236–239
11. Cole AJ, David AE, Wang JX, Galban CJ, Yang VC (2011) Magnetic brain tumor targeting and biodistribution of long-circulating PEG-modified, cross-linked starch-coated iron oxide nanoparticles. Biomaterials 32:6291–6301
12. Guo JW et al (2011) Aptamer-functionalized PEG-PLGA nanoparticles for enhanced anti-glioma drug delivery. Biomaterials 32:8010–8020
13. Ferlay J, Shin HR, Bray F, Forman D, Mathers C, Parkin DM (2010) Estimates of world wide burden of cancer in 2008: GLOBOCAN 2008. Int J Cancer 127:2893–2917

14. Agarwal A et al (2011) Dual-mode imaging with radiolabeled gold nanorods. J Biomed Opt 16(051307):1–7
15. Tuan V-D et al (2013) Plasmonic nanoprobes: from chemical sensing to medical diagnostics and therapy. Nanoscale 5:10127–10140
16. Ashton JR et al (2014) Dual-energy micro-CT functional imaging of primary lung cancer in mice using gold and iodine nanoparticle contrast agents: a validation study. PLoS ONE 9 (e88129):1–14
17. Toy R, Peiris PM, Ghaghada KB, Karathanasis E (2014) Shaping cancer nanomedicine: the effect of particle shape on the in vivo journey of nanoparticles. Nanomedicine 9:121–134
18. Yuan H, Fales AM, Khoury CG, Liu J, Vo-Dinh T (2013) Spectral characterization and intracellular detection of surface-enhanced Raman scattering (SERS)-encoded plasmonic gold nanostars. J Raman Spectrosc 44:234–239
19. Liu Y, Yuan H, Kersey FR, Register JK, Parrott MC, Vo-Dinh T (2015) Plasmonic gold nanostars for multi-modality sensing and diagnostics. Sensors 15:3706–3720
20. Sun XL et al (2014) Chelator-free Cu-64-integrated gold nanomaterials for positron emission tomography imaging guided photothermal cancer therapy. ACS Nano 8:8438–8446
21. Seo JW et al (2015) Self-assembled 20-nm Cu-64-micelles enhance accumulation in rat glioblastoma. J Controlled Release 220:51–60
22. Gao N et al (2014) Shape-dependent two-photon photoluminescence of single gold nanoparticles. J Phys Chem C 118:13904–13911
23. Black KCL et al (2014) Radioactive Au-198-hoped nanostructures with different shapes for in vivo analyses of their biodistribution, tumor uptake, and intratumoral distribution. ACS Nano 8:4385–4394

Chapter 4
GNS Toxicity Investigation

4.1 Background

GNS toxicity study evaluating biocompatibility is crucial for their biomedical applications and future clinical translation [1, 2]. Previous studies with various animal models including mice, rats and dogs have demonstrated that gold nanoparticles with a spherical shape are biocompatible [3–8]. However, few studies have reported toxicity evaluation of nonspherical gold nanoparticles (like star-shape). Herein, we performed both in vitro and in vivo toxicity studies to investigate potential deleterious effect of PEGylated GNS nanoparticles after IV administration.

4.2 In Vitro Toxicity Study

To begin assessing the potential toxicity of the GNS probe, the WST-8 assay, a colorimetric assay for cell viability and proliferation evaluation, was used on neural stem cells (NSCs) to quantify the proportion of cells that remained viable after exposure to various concentrations of GNS nanoparticles. The tetrazolium salt in the WST-8 assay can be reduced to a soluble purple formazan by a reducing agent, NADH in cells. The produced formazan was measured with absorbance at 450 nm. NSCs were plated one day prior to GNS administration. After the addition of GNS in various concentrations, cells were incubated at 37° with 5% CO_2 in Stem Cell Technologies's NeuroCult™ Proliferation Kit with 10% proliferation supplement and 2% EGF for three days. On the third day, 10 µl of CCK-8 solution was added to the cell culture media to incubate for an additional 6 h. The plate was then transferred to the TECAN Infinite Pro Microplate Reader (Switzerland) to determine the absorbance reading of the treated and untreated cells. The absorbance was

Y. Liu, *Multifunctional Gold Nanostars for Cancer Theranostics*, Springer Theses,
https://doi.org/10.1007/978-3-319-74920-4_4

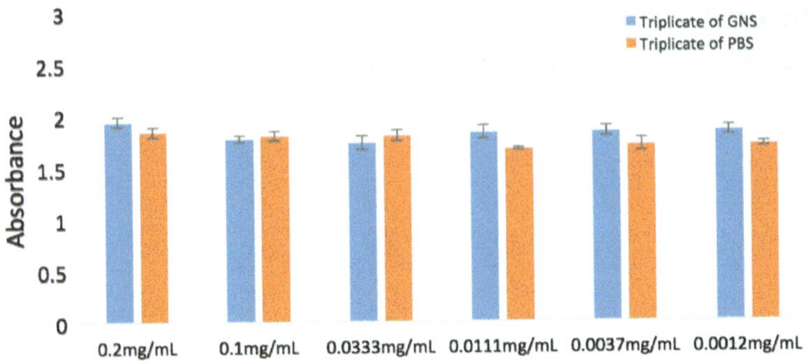

Fig. 4.1 The WST-8 assay demonstrates the biocompatibility of GNS administered on mouse NSCs after a three-day incubation of GNS at various concentrations. Cell viability was not significantly altered due to application of GNS at dose up to 0.2 mg/ml. The error bar is shown as the standard deviation

measured at 450 nm with a reference wavelength at 650 nm. As shown in Fig. 4.1, no statistical difference was found between the treatment group and control group.

4.3 In Vivo Toxicity Study

After in vitro toxicity study, we performed a comprehensive in vivo study including mice behavior observation, body weight monitoring, blood chemistry test and H&E histopathology examination to assess the potential toxicity of PEGylated GNS nanoparticles 6 months after IV administration. Statistical analysis on body weight was performed with a mixed-model ANOVA to assess the effect of gold nanoparticle dosage on body weight over time. One-way ANOVA statistical analysis was performed to compare the effect of GNS dosage (PBS, 20, and 80 mg/kg) on different aspects of blood chemistry test items including BUN, creatinine, calcium, protein, albumin, globulin, glucose, cholesterol, ALT (GPT), AST (GOT), ALP and bilirubin. For all comparisons, the data was considered statistically significant for $P < 0.05$ (95% confidence interval).

During the 6-month toxicity study, all mice were carefully monitored and found to not exhibit stress or any other abnormal behavior. Body weight was used as a guideline for assessing mice's general health condition and was monitored weekly. Experimental results (Fig. 4.2) show no difference with statistical significance on body weight between the control group and groups receiving GNS doses up to 80 mg/kg by mixed-model ANOVA analysis (criteria $P = 0.05$, 95% confidence).

Mice were sacrificed 6 months after GNS IV injection and plasma was harvested for blood chemistry test. ANOVA statistical analyses were performed and no statistically significant difference on blood chemistry test results between the control group and study groups receiving 20 mg/kg or 80 mg/kg GNS dose was observed

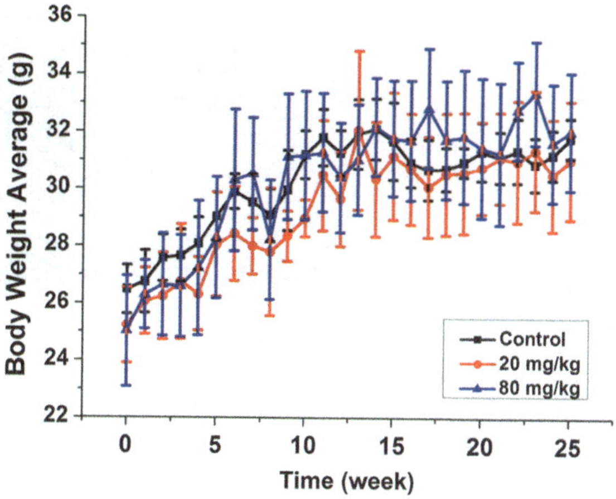

Fig. 4.2 Body weight was monitored once per week after GNS injection for 6 months. Mixed-model ANOVA was used to perform statistical analysis and results show that there is no difference between control group and groups with 20 and 80 mg/kg dose. The result is considered statistically significant for $P < 0.05$ (95% confidence interval). Each group contained 4 mice and error bar was shown as the standard deviation

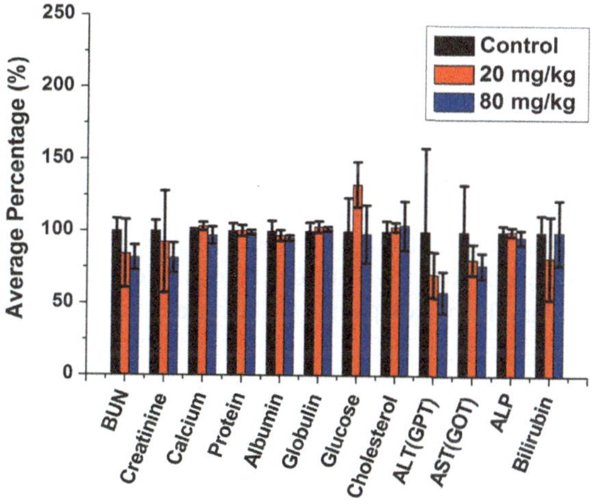

Fig. 4.3 Mice were sacrificed 6 months after GNS injection and blood was harvested for blood chemistry test. One-way ANOVA was used to perform statistical analysis and results show that there is no difference between control group and groups with 20 and 80 mg/kg dose. The result is considered statistically significant for $P < 0.05$ (95% confidence interval). Each group contained 4 mice and error bar was shown as the standard deviation

(Fig. 4.3). Blood chemistry test results showed preservation of liver and kidney function after GNS IV injection. H&E histopathology examination of brain, heart, liver, kidney and spleen was also performed (Fig. 4.4). No remarkable tissue

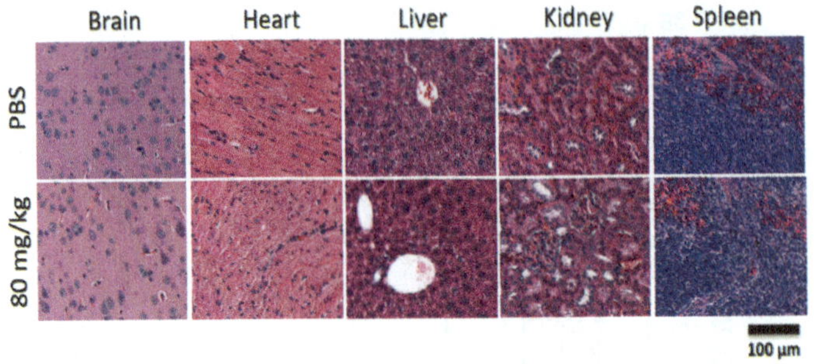

Fig. 4.4 H&E histopathology examination of various organs of mice 6 months after 80 mg/kg GNS or PBS (control) IV injection through tail vein. No tissue structure changes related with GNS toxicity was identified. Scale bar, 100 µm

structure changes indicating GNS-related toxicity up to 80 mg/kg dose were identified by a veterinary pathologist.

After in vivo long-term toxicity, we also performed a short-term study with high dose (500 mg/kg), 10–100 folds higher than typically in vivo used dosage, to investigate potential acute toxicity. None of 5 mice died one week after IV injection.

4.4 Summary

Toxicity study results including in vitro cell test, 1-week acute toxicity study with high dose, 6-month chronic evaluation of behavior, body weight monitoring, blood chemistry analysis, and H&E pathology examination showed that our PEGylated GNS nanoparticles had no deleterious effects after IV injection with dose up to 80 mg/kg. The biocompatible GNS warrants further evaluation for future clinical translation.

References

1. Perazella MA (2009) Current status of gadolinium toxicity in patients with kidney disease. Clin J Am Soc Nephrol 4:461–469
2. Muldoon LL, Sandor M, Pinkston KE, Neuwelt EA (2005) Imaging, distribution, and toxicity of superparamagnetic iron oxide magnetic resonance nanoparticles in the rat brain and intracerebral tumor. Neurosurgery 57:785–796
3. Khlebtsov N, Dykman L (2011) Biodistribution and toxicity of engineered gold nanoparticles: a review of in vitro and in vivo studies. Chem Soc Rev 40:1647–1671
4. Gad SC, Sharp KL, Montgomery C, Payne JD, Goodrich GP (2012) Evaluation of the toxicity of intravenous delivery of Auroshell particles (gold-silica nanoshells). Int J Toxicol 31: 584–594

5. Yen HJ, Hsu SH, Tsai CL (2009) Cytotoxicity and immunological response of gold and silver nanoparticles of different sizes. Small 5:1553–1561
6. Cho WS et al (2009) Acute toxicity and pharmacokinetics of 13 nm-sized PEG-coated gold nanoparticles. Toxicol Appl Pharmacol 236:16–24
7. Shukla R, Bansal V, Chaudhary M, Basu A, Bhonde RR, Sastry M (2005) Biocompatibility of gold nanoparticles and their endocytotic fate inside the cellular compartment: a microscopic overview. Langmuir 21:10644–10654
8. Connor EE, Mwamuka J, Gole A, Murphy CJ, Wyatt MD (2005) Gold nanoparticles are taken up by human cells but do not cause acute cytotoxicity. Small 1:325–327

Chapter 5
Sensitive Brain Tumor Detection Using GNS Nanoprobe

5.1 Background

Malignant brain tumors are one of the most lethal, heterogeneous, and difficult cancer forms to treat [1–5]. For the most common malignant glioblastoma (GBM), an advanced grade IV astrocytoma, the median overall survival (OS) is only 15 months with current standard therapies [6]. Tumor size has been found to be an independent prognostic factor and patients having small tumors when diagnosed have been reported to show longer OS and better prognosis [7, 8]. Therefore, novel methods for sensitive brain tumor detection offer an important prospect for significant improvement of patients' outcomes. The current routine brain tumor diagnosis method is magnetic resonance imaging (MRI) with and without contrast agent enhancement [9, 10]. The size limit for brain tumor detection with MRI scan has been found to be 3–5 mm and it is highly challenging for MRI to detect a brain tumor less than 1 mm [11, 12]. Positron emission tomography (PET) provides a sensitive 3D imaging modality (several orders of magnitude more sensitive than MRI for contrast agent detection), and has been applied in clinics for cancer detection using fluorodeoxyglucose ([^{18}F]FDG) radiotracer due to higher glucose uptake in tumors than in normal tissues [13–19]. However, PET imaging using [^{18}F]FDG is not ideal for brain tumor detection due to the high glucose uptake in normal brain tissue, resulting in low tumor-to-normal (T/N) ratio and high false positive rates [20]. To date, no study has demonstrated the feasibility of submillimeter brain tumor detection using noninvasive diagnostic modalities. Sensitive submillimeter brain tumor detection is still an unmet clinical need of critical importance. Our GNS-based nanoprobe for PET imaging has potential to provide a novel method for early brain tumor detection.

Y. Liu, *Multifunctional Gold Nanostars for Cancer Theranostics*, Springer Theses,
https://doi.org/10.1007/978-3-319-74920-4_5

5.2 Serial PET Imaging for GNS Tumor Uptake Study

For in vivo brain tumor detection study, 30 nm PEGylated GNS nanoprobes were labeled with ^{124}I radioisotope for PET imaging. The orthotopic brain tumor mouse model was generated by intracranial injection of mouse neural stem cells with isocitrate dehydrogenase-1 (IDH-1), p53, and platelet derived growth factor subunit B (PDGFB) gene mutations. The mice with injected neoplastic cells were monitored for three months. PET/CT scans were performed using an animal PET/CT scanner (Siemens Inveon PET/CT, Siemens Medical Systems, Knoxville TN USA). Mice with brain tumor were IV injected with ^{124}I-labeled GNS (500 μg in gold mass and 100 μCi in ^{124}I radioactivity) through tail vein. 10-min PET scans were immediately started after IV injection and then CT scan was followed to show anatomical structure. Follow-up scans were started 4, 24, 48, and 120 h post injection. In addition, one mouse with brain tumor was used to compare [^{18}F]FDG and ^{124}I labeled GNS for brain tumor detection under PET scan. For comparison studies, 100 μCi of [^{18}F]FDG were IV injected through tail vein and PET/CT scans were performed 1 h post injection. One day after [^{18}F]FDG injection, ^{124}I labeled GNS nanoprobes (500 μg in gold mass and 100 μCi in radioactivity) were IV injected through tail vein. PET and CT scans were performed 24 and 48 h after GNS IV injection. Mouse brain was harvested after PET/CT scan and fixed with formalin for histopathology and optical imaging. TPL imaging was performed using an Olympus FV 1000 multiphon microscope with a Ti: Sapphire fsec laser and a 25x/1.05 NA water objective. The laser is tunable in the range of 680–1080 nm and has an 80 MHz repetition rate and a 140-fsec pulse width. Haematoxylin and eosin (H&E) imaging was performed by using a Zeiss Axio Imager wide-field microscope coupled with an Axiocam 506 color camera.

Longitudinal PET/CT scans were performed to investigate the GNS accumulation kinetics in brain tumors. Figure 5.1 depicts dynamic PET/CT scans obtained at 10 min, 4, 24, 48 and 120 h following IV injection of ^{124}I labeled GNS nanoprobes. Starting at the 24 h time point, the tumor uptake was higher than that in normal brain with the contrast ratio increasing as time progressed. The tumor-to-normal ratio (T/N) is 1.0, 2.5, 3.8, 7 and 7.8 at 10 min, 4, 24, 48 and 120 h, respectively.

5.3 Comparison of GNS and FDG Radiotracer for Brain Tumor Detection

PET imaging using ^{124}I-GNS was compared with that using [^{18}F]FDG for brain tumor detection as shown in Fig. 5.2a. The [^{18}F]FDG has a much faster clearance speed than ^{124}I-GNS, and therefore, the PET scan was performed one hour after [^{18}F]FDG IV injection while the PET scan was performed 48 h after I-124 labeled

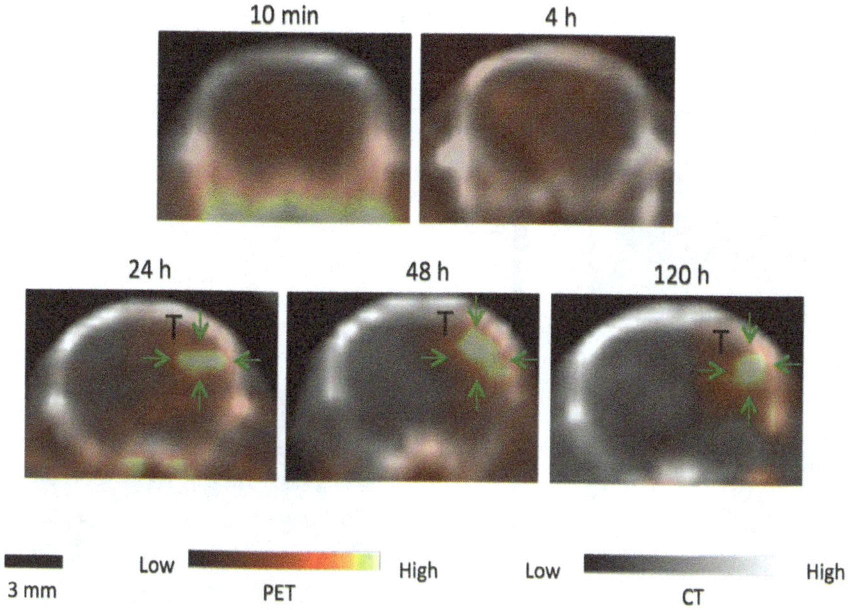

Fig. 5.1 PET/CT imaging of ^{124}I-GNS nanoprobe accumulation in the brain tumor at five time points. The tumor (T; green arrows) has higher ^{124}I-GNS uptake than the contralateral normal brain 24 h after IV injection. The tumor uptake (average) is measured to be 0.76%ID/g (10 min), 1.5%ID/g (4 h), 1.3%ID/g (24 h), 0.91%ID/g (48 h) and 0.69%ID/g (120 h). The tumor-to-normal ratio (T/N) is calculated to be 1.0 (10 min), 2.5 (4 h), 3.8 (24 h), 7.0 (48 h) and 7.8 (120 h)

GNS IV injection. The tumor region had higher GNS uptake than that in normal brain (T/N = 4.0); on the other hand, the uptake of ^{18}F-FDG in tumor and normal brain was almost the same. H&E examination was performed to confirm the brain tumor area (Fig. 5.2b). High-resolution TPL imaging (Fig. 5.2c) confirmed the GNS accumulation (white spots) inside the tumor, consistent with the PET imaging results.

5.4 Tumor Detection Limit Using GNS Nanoprobe

We also sought to determine the size limit of brain tumor detection with our ^{124}I-labeled GNS nanoprobe. As shown in Fig. 5.3a, a small "hot spot" was identified in the PET/CT scan 48 h following GNS IV injection. The T/N ratio is measured to be 4.7 and the tumor uptake was found to be much higher than surrounding normal brain tissue. After PET/CT scan, the mouse was sacrificed and the brain was harvested for histopathology examination. The mouse brain was

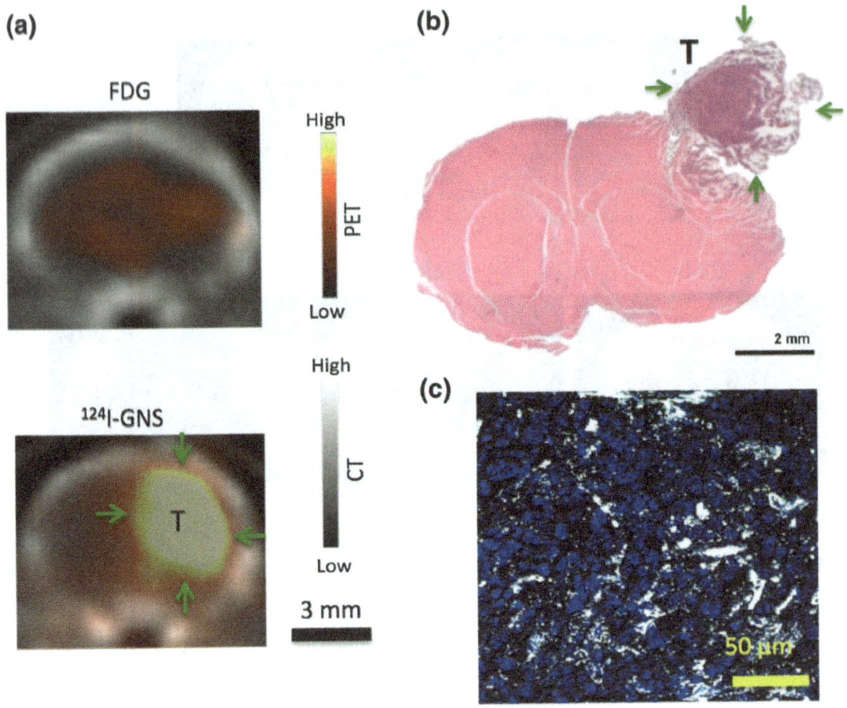

Fig. 5.2 **a** Comparison of ^{18}F-FDG and ^{124}I-GNS for brain tumor detection by PET imaging in the same brain tumor-bearing mouse. The average tumor uptake of ^{124}I-GNS nanoprobes is measured to be 7.2%ID/g and the T/N ratio is calculated to be 4.0 while the T/N ratio for FDG is 1.1. **b** H&E histopathology examination of the brain tumor detected by PET scan as shown in **a**. Green arrows are used to show tumor (T). **c** TPL imaging shows GNS (white spots) are inside brain tumor detected from PET scan **a**. The section is stained with DAPI and nuclei are shown as blue color. The histopathology image represents a 5 μm tissue section. Tumor came off brain during brain harvest process and image was reconstructed to combine tumor with brain

sectioned slice by slice at the tumor location guided by the PET/CT scan. H&E staining was performed on those sections to mark the tumor boundary and measure the tumor size. As shown in Fig. 5.3b, c, the small tumor detected by the PET/CT scan had a size less than 0.5 mm in diameter. A series of brain sections with H&E staining confirmed that the tumor's size was less than 0.5 mm in all three dimensions. In addition, TPL imaging was also performed to confirm the presence of GNS inside the detected 0.5 mm brain tumor. As shown in Fig. 5.3d, GNS (white spots) nanoprobes were indeed inside the 0.5 mm brain tumor detected by PET/CT imaging.

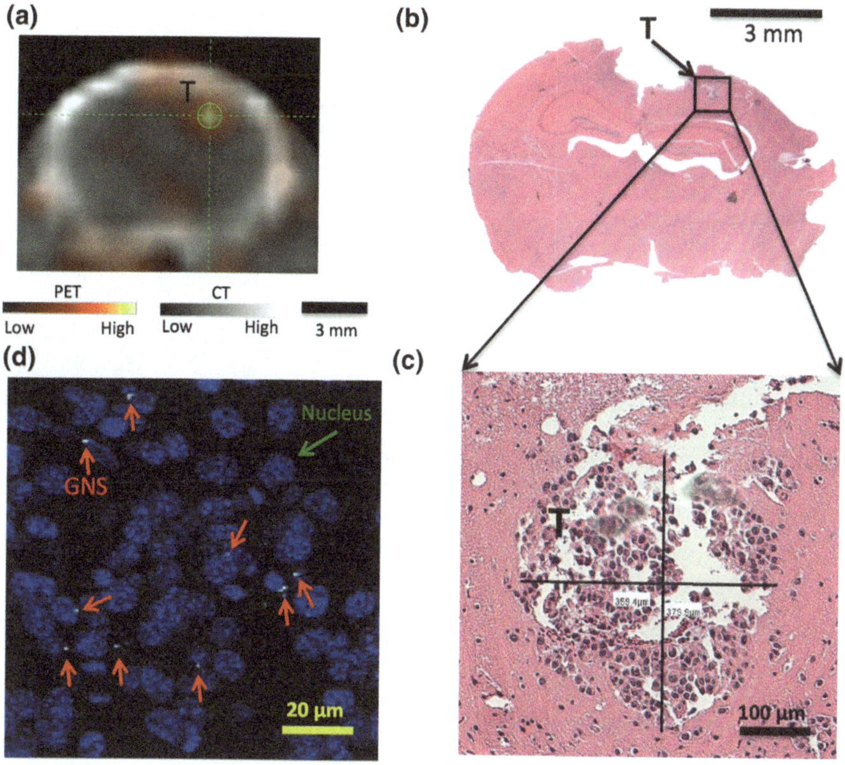

Fig. 5.3 **a** 0.5 mm brain tumor (T) identified from PET/CT scan 48 h after IV injection. The average tumor uptake is measured to be 0.66%ID/g and the T/N is measured to be 4.7. **b** H&E histopathology examination confirms the identified brain tumor from PET/CT scan **a**. **c** The identified tumor size is less than 0.5 mm measured from H&E imaging. **d** TPL imaging showed the GNS (white spots, marked by red arrow) were inside the identified 0.5 mm brain tumor. The tumor cell nuclei were stained with DAPI and shown in blue color. Certain GNS nanoprobes are found to be near tumor nuclei

5.5 Study of GNS Uptake in Brain Tumor

Furthermore, we performed TEM imaging to identify the subcellular location of GNS nanoprobes at a nanometer resolution level after IV injection. As shown in Fig. 5.4a, b, GNS nanoprobes were found to leak through the vasculature disrupted by tumor and tended to accumulate in the tumor's interstitial space. Figure 5.4c, d show that GNS nanoprobes diffused through extracellular space and were endocytosized inside brain tumor cells by endosomes. On the contrary, as shown in Fig. 5.5, in the normal brain tissue, GNS nanoprobes were blocked by the intact blood-brain-barrier and found near the inner blood vessel wall.

Using both optical and TEM imaging, we demonstrated that the GNS nanoprobes not only crossed a disrupted blood-brain-barrier in brain tumors and

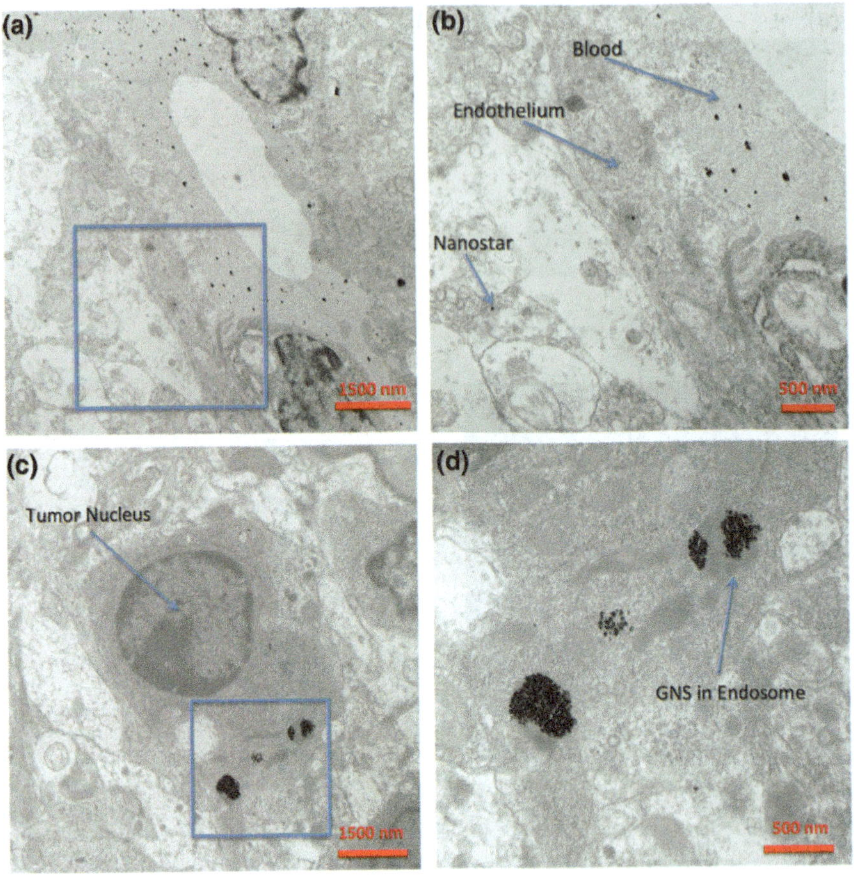

Fig. 5.4 **a**, **b** TEM imaging of GNS in the extracellular space of the tumor region. **c**, **d** TEM imaging of GNS in endosomes inside brain tumor cells

penetrated through the tumor interstitial space, but were also endocytosed into tumor cells following IV tail vein injection. ^{124}I has a longer half-life (4.18 days) than ^{64}Cu (12.7 h) or ^{18}F (1.8 h), which makes it more suitable for in vivo nanoparticle tracking with PET imaging over a long time period. The observation of GNS accumulation in brain tumors has important implications for future therapeutic agents delivery. The accumulation effect could be further enhanced when GNS nanoprobes are administered with vasodilating agents, which have been shown to magnify the EPR effect [21]. Moreover, from a chemical perspective, the I-Au chemical bond enables a simple radiolabeling process of ^{124}I without the need of chelators or complicated radiochemistry reactions. The simple radiolabeling process of gold nanoparticles with ^{124}I would be readily adaptable to future clinical PET applications. Furthermore, the large surface area of gold nanostars can also be used

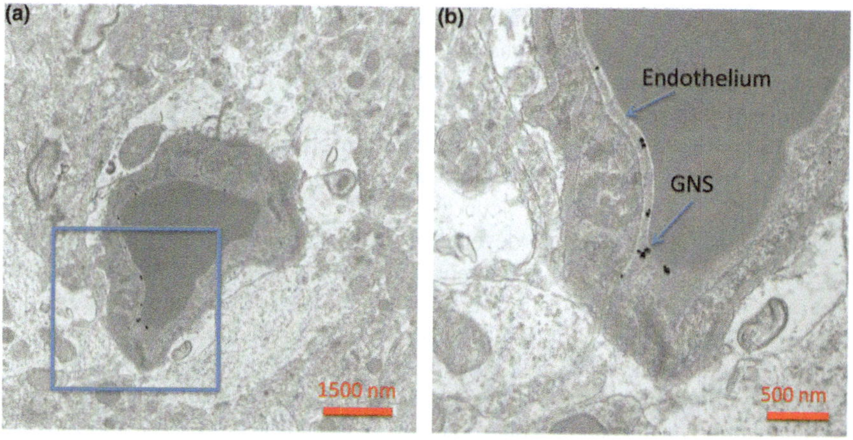

Fig. 5.5 **a**, **b** TEM imaging of GNS identified inside the normal brain vasculature. GNS nanoprobes were found to be blocked by the intact blood-brain-barrier and near the inner endothelium wall

as an efficient carrier of chemically similar therapeutic radioisotopes such as ^{131}I or ^{211}At for targeted radionuclide cancer therapy.

It is of interest to demonstrate selective accumulation of these GNS nanoprobes in the brain tumor after IV injection. The BBB in normal brain has a pore size less than 1 nm, which limits passive flow of large particles. BBB permeability is increased as a result of malignant brain tumor pathogenesis, which includes neovascularization [22]. Tumors as small as 200 μm rely on this neovasculature to provide nutrition, oxygen supply, as well as a means of metabolic waste product clearance. These blood vessels often have different characteristics from normal blood vessels, including defective endothelial cells with wide fenestrations and an absence of the smooth muscle layer. A lack of pinocytotic vesicles within the cerebral capillary endothelial cells lead to inefficiency of transcytosis [23]. Therefore, we propose that the major reason for PEGylated GNS accumulation in brain tumor is passive targeting via the EPR effect due to the leaky or abnormally formed vasculature caused by the brain tumor. Our GNS nanoprobes functionalized with PEG polymer were designed with suitable particle size and surface properties to leak through those gaps in the tumor vasculature and show selective accumulation in brain tumor but not into surrounding normal brain tissues, which have intact BBB. Consistent with our observations, it has been reported that nanoparticles up to 114 nm with dense PEG coating can diffuse through the brain extracellular space; moreover, it was estimated that the brain extracellular space has more than 25% of pores with size larger than 100 nm [24]. The GNS nanoprobes observed inside tumor cells are presumed to have penetrated through those pores and enter tumor cells by endocytosis. Our results further support that the EPR effect is an important mechanism for cancer targeting.

By using our novel GNS nanoprobes combined with highly sensitive PET imaging, we have demonstrated the feasibility of detecting brain tumors as small as 0.5 mm, which is one order of magnitude less than current clinical imaging modalities [8]. In addition, compared to MRI, which has high background signal in normal brain tissue and relatively low contrast discrimination due to compositions with high protons like water and fat, our method using radiolabeled GNS shows minimal background in normal brain tissue and a high tumor-to-normal ratio. To the best of our knowledge, the developed GNS nanoprobe for PET scan provides the most sensitive means of brain tumor detection reported so far. It is well known that early detection of the smallest brain tumor volume (or cell number) at the time of cancer diagnosis has significant impact on the therapeutic efficacy and overall patient prognosis. The tumor volume is directly proportional to the 3rd power of the diameter. Therefore, tumor volume is three orders of magnitude smaller if the detection limit in diameter is one order of magnitude less. The developed GNS nanoprobes with high tumor-to-normal ratio and submillimeter brain tumor detection capability is of great importance to improve brain tumor patients' outcome. In addition to whole body scan using PET/CT, the developed multifunctional GNS nanoprobe can also be imaged using optical imaging and electron microscopy without the need of extra dye labeling or staining, which provides a superior tool to investigate how nanoparticles can be used for brain tumor drug delivery through BBB.

5.6 Summary

We have developed a multifunctional ^{124}I-labeled GNS nanoprobe for sensitive in vivo brain tumor detection using PET imaging as well as for high-resolution nanoparticle tracking using both optical and TEM imaging. PET imaging using ^{124}I-GNS achieved ultrasensitive brain tumor detection down to 0.5 mm in size in a laboratory mouse model. After IV injection, GNS nanoparticles were found inside brain tumor cells by using both TPL and TEM imaging. The GNS nanoprobes permeated the disrupted brain tumor vasculature and subsequently entered tumor cells by endocytosis. Therefore, GNS nanoparticles can be used as a promising nanoplatform for future early brain tumor detection and therapeutic agent delivery.

References

1. Tagami T, Taki M, Ozeki T (2016) Nanoparticulate drug delivery systems to overcome the blood-brain barrier. In: Lu ZR, Sakuma SN (eds) Nanomaterials in pharmacology. Humana Press Inc., New York City
2. Han L et al (2016) Increased nanoparticle delivery to brain tumors by autocatalytic priming for improved treatment and imaging. ACS Nano 10:4209–4218

3. Rutka JT, Kim B, Etame A, Diaz RJ (2014) Nanosurgical resection of malignant brain tumors: Beyond the cutting edge. ACS Nano 8:9716–9722
4. Kircher MF et al (2012) A brain tumor molecular imaging strategy using a new triple-modality MRI-photoacoustic-Raman nanoparticle. Nat Med 18:829–834
5. Kohler BA et al (2011) Annual report to the nation on the status of cancer, 1975–2007, featuring tumors of the brain and other nervous system. J Nat Cancer Inst 103:714–736
6. Krex D et al (2007) Long-term survival with glioblastoma multiforme. Brain 130:2596–2606
7. Mohammadi AM et al (2016) Impact of the radiosurgery prescription dose on the local control of small (2 cm or smaller) brain metastases. J Neurosurg 126(3):1–9
8. Dempsey MF, Condon BR, Hadley DM (2005) Measurement of tumor "size" in recurrent malignant glioma: 1D, 2D, or 3D? AJNR Am J Neuroradiol 26:770–776
9. Schellinger PD, Meinck HM, Thron A (1999) Diagnostic accuracy of MRI compared to CCT in patients with brain metastases. J Neurol Oncololgy 44:275–281
10. Sze G, Milano E, Johnson C, Heier L (1990) Detection of brain metastases: comparison of contrast-enhanced MR with unenhanced MR and enhanced CT. AJNR Am J Neuroradiol 11:785–791
11. Erdi YE (2012) Limits of tumor detectability in nuclear medicine and PET. Mol Imaging Radionuclide Ther 21:23–28
12. Yuh WT, Tali ET, Nguyen HD, Simonson TM, Mayr NA, Fisher DJ (1995) The effect of contrast dose, imaging time, and lesion size in the MR detection of intracerebral metastasis. Am J Neuroradiol 16:373–380
13. Fletcher JW et al (2008) Recommendations on the use of F-18-FDG PET in oncology. J Nucl Med 49:480–508
14. Rohren EM, Turkington TG, Coleman RE (2004) Clinical applications of PET in oncology. Radiology 231:305–332
15. Phelps ME (2000) PET: The merging of biology and imaging into molecular imaging. J Nucl Med 41:661–681
16. Bradley J et al (2004) Impact of FDG-PET on radiation therapy volume delineation in non-small-cell lung cancer. Int J Radiat Oncol Biol Phys 59:78–86
17. Boellaard R et al (2010) FDG PET and PET/CT: EANM procedure guidelines for tumour PET imaging: version 1.0. Eur J Nucl Med Mol Imaging 37:181–200
18. Antoch G et al (2003) Non-small cell lung cancer: Dual-modality PET/CT in preoperative staging. Radiology 229:526–533
19. MacManus M et al (2009) Use of PET and PET/CT for radiation therapy planning: IAEA expert report 2006–2007. Radiother Oncol 91:85–94
20. Shreve PD, Anzai Y, Wahl RL (1999) Pitfalls in oncologic diagnosis with FDG PET imaging: physiologic and benign variants. Radiographics 19:61–77
21. Maeda H, Tsukigawa K, Fang J (2016) A retrospective 30 years after discovery of the enhanced permeability and retention effect of solid tumors: next-generation chemotherapeutics and photodynamic therapy—problems, solutions, and prospects. Microcirculation 23:173–182
22. Jain KK (2012) Nanobiotechnology-based strategies for crossing the blood-brain barrier. Nanomedicine 7:1225–1233
23. Lesniak MS, Brem H (2004) Targeted therapy for brain tumours. Nat Rev Drug Discovery 3:499–508
24. Nance EA et al (2012) A dense poly(ethylene glycol) coating improves penetration of large polymeric nanoparticles within brain tissue. Sci Transl Med 4(8149):1–8

Chapter 6
Photoimmunotherapy for Cancer Metastasis Treatment

6.1 Background

Cancer metastasis has caused more than 90% of cancer deaths [1]. When the cancer cells have spread throughout the body, limited therapeutic options, such as systemic chemotherapy, are often not quite effective to control the disease [2–8]. Immunotherapies have emerged as one of the most-promising modalities to treat cancer and have involved the use of drugs that target immune checkpoints [9–11]. The programmed death-ligand 1 (PD-L1), a protein overexpressed on cancer cell membrane, contributes to the suppression of the immune system [12–16]. PD-L1 ligand binds to its receptor, PD-1, found on activated T cells, B cells, and myeloid cells, to modulate T cell functions [16–20]. The therapeutic anti-PD-L1 antibody is designed to block the PD-L1/PD-1 interaction and reverse tumor-mediated immunosuppression [19, 21]. Blocking the PD-L1/PD-1 axis has been shown to be highly beneficial in clinical trials [22–24]. Although PD-L1/PD1 immunotherapy has emerged as one of the most-promising modalities to treat systemic cancer, they work only for a limited number of patients and can become ineffective with time. Broadening and stabilizing therapeutic effects of PD-L1/PD-1 inhibitors could potentially address the challenge of successfully treating metastatic cancers.

The plasmonic GNS enhanced photothermal therapy has potential to generate synergistic effect with immunotherapy by ablating primary tumor cells to release tumor-specific neoantigen and trigger an immune response. The immune system can be triggered by several temperature-induced mechanisms, such as antigen presentation mediated by heat shock proteins (HSPs) or lymphocytes migration to lymphoid organs, the staging ground for immune defense. Therefore, immunotherapies could potentially benefit from targeted thermal therapies and generate synergistic effect to treat cancer more efficiently.

Y. Liu, *Multifunctional Gold Nanostars for Cancer Theranostics*, Springer Theses,
https://doi.org/10.1007/978-3-319-74920-4_6

6.2 Window Chamber Imaging and In Vivo Photothermal Study

First, we used window chamber model and two-photon microscopy to confirm GNS accumulation in the tumor developed with the bladder cancer cells. As shown in Fig. 6.1a, the tumor under window chamber has clear GNS accumulation (black color) one day after IV injection through tail vein. The black color is due to strong light absorption of accumulated GNS nanoparticles in tumor. The tumor size is approximately 2 mm, which shows that even a small tumor exhibit the EPR effect associated with leaky vasculature. Figure 6.1b shows the high-resolution imaging of GNS nanoprobes in tumor through window chamber using TPL. GNS

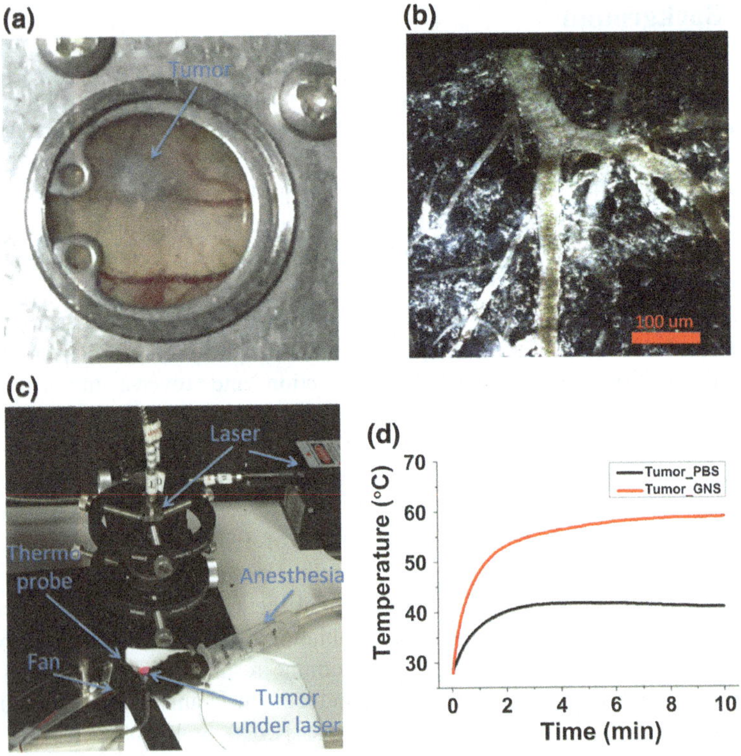

Fig. 6.1 **a** Tumor with GNS selective accumulation (black color) under window chamber (8 mm in diameter). The tumor shows black color because GNS have strong light absorption. **b** Two-photon photoluminescence (TPL) imaging of GNS nanoparticles in tumor under window chamber, due to their unique plasmonic properties, GNS nanoprobes emit strong TPL emission allowing direct particle visualization under multiphoton microscopy. The GNS are shown as white color under TPL imaging. The scale bar is 100 μm. **c** Laser treatment setup developed for photothermal ablation. **d** Tumor center temperature measured with mini-invasive thermoprobe for the mouse with GNS or PBS injection under 0.6 W/cm^2 808-nm laser irradiation

nanoprobes have exceptionally high TPL cross section and show bright white color. GNS nanoparticles are found in both tumor vasculature and interstitial space. In addition, a setup for photothermal treatment was built in lab (Fig. 6.1c) and a mini-invasive thermoprobe was used to measure the tumor center temperature under laser irradiation for the mouse with GNS nanoprobes or PBS (control) IV injection through tail vein. It can be seen that under 0.6 W/cm^2 808-nm laser irradiation, tumor with GNS accumulation reaches ablation temperature 58 °C while tumor without GNS just 42 °C (Fig. 6.1d). Experimental results demonstrate that GNS, with selective accumulation in tumor and high photon-to-heat conversion efficiency, can be used as a powerful thermal transducer for photothermal therapy to ablate tumor.

6.3 Long Term Survival and Rechallenge Study for Photoimmunotherapy

To evaluate the potential synergistic effect between plasmonic GNS-enabled PTT and immune checkpoint inhibitor immunotherapy (anti-PD-L1), MB49 bladder cancer cells were implanted in both right and left flanks (250,000/100,000, respectively) of C57BL/6 mice [25]. These aggressive cancer cells took only one week to grow to 100–200 mm^3 (right flank) and 50–100 mm^3 (left flank) in tumor size. At this time point (Day 0), mice were intravenously injected with 30 nm PEGylated GNS at a dose of 80 mg/kg in 100 μl of PBS solution through tail vein. One day after GNS injection, an 808-nm laser (0.6 W/cm^2) was used to irradiate right flank tumor for 10 min with anesthesia using isoflurane. Immediately following laser treatment, mice were intraperitoneal (IP) injected with anti-PD-L1 antibodies (200 μg per mouse every 3 days until the end of the study). Tumor size and body weight were measured every three times every week. Humane endpoint euthanasia was performed if the mice show adverse reactions to treatments, lose body weight greater than 15%, exhibit a single tumor volume greater than 1000 mm^3 (4% body weight) or total tumor volume is greater than 1500 mm^3 (6% body weight).

As shown in Fig. 6.2a, the primary tumors (right flank) treated with laser in the treatment group shrank starting on Day 8. Two mice (1 and 3) exhibited no measurable tumors from day 26 until day 49. Two mice (2 and 4) had tumor sizes that decreased but were sacrificed due to the large tumors on the other side. One mouse (5) had tumor sizes that decreased first but then increased after approximately one week. Not only is there an immediate therapeutic effect at the tumor site treated with NIR laser, but also the PTT treatment results in a general activation of the immune system, as evidenced by the fact that distant tumors without laser treatment also show therapeutic effect. As shown in Fig. 6.2b, for the distant tumor (used as cancer metastasis model), 2 mice (1 and 3) were tumor-free between Day 25 and Day 33. Both laser treated and untreated tumors of mouse 1 in the treatment group

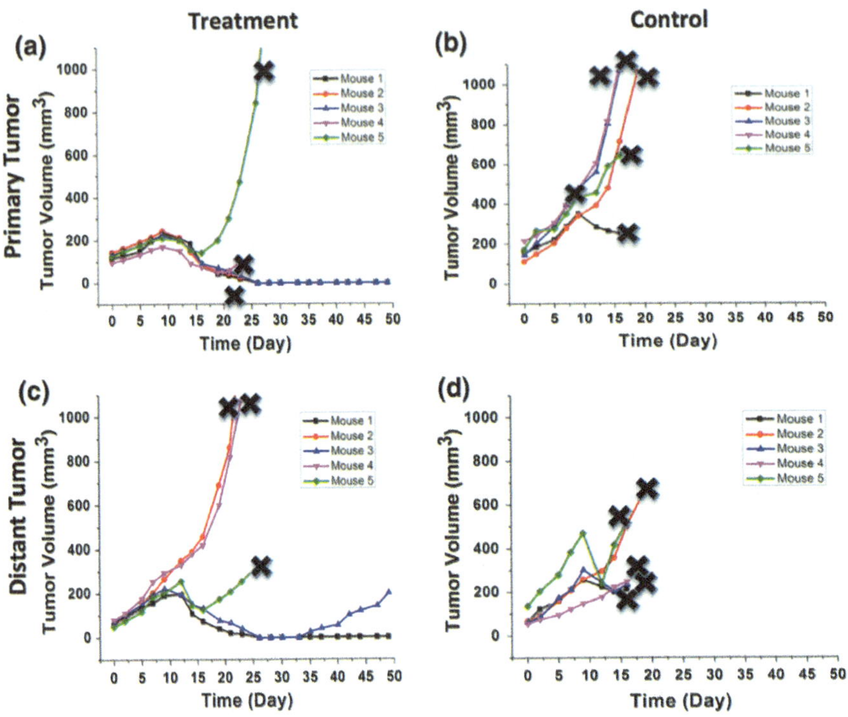

Fig. 6.2 Primary tumor size change profile for the mice with photoimmunotherapy treatment (**a**) and blank control (**b**). Distant tumor size change profile for the mice with photoimmuotherapy treatment (**c**) and blank control (**d**). The laser irradiation of photothermal therapy was performed only on the primary tumor. The line stopped (× sign in black color) if the mouse was sacrificed due to large tumor or ulceration. (Adapted from Ref. [25])

disappear after Day 36. In the contrary, primary and distant tumors in the blank control group exhibited rapid and significant growth until all mice were sacrificed.

Figure 6.3 illustrates the Kaplan-Meier (K-M) overall survival curve, which shows significant survival improvement of the photoimmunotherapy group over the control groups. At the end of 49 days, the survival rate for photoimmunotherapy group is 40 and 0% for all control groups including the anti-PD-L1 antibody therapy alone group. The anti-PD-L1 therapy alone did show therapeutic effect compared to the blank control group receiving no treatment. However, the photoimmunotherapy group had an improved therapeutic effect compared with the anti-PD-L1 antibody alone group. Only the photoimmunotherapy group had one survival mouse without tumor 3 months after treatment, indicating a complete cure.

In addition, we performed a rechallenge study by investigating whether the generated immune response has a memory effect and can last for a long time. The two mice that were tumor free for 4 months were injected with 250,000 MB49 bladder cancer cells under dorsal skin. No tumors developed 2 months after injection (Fig. 6.4), indicating the existence of a memorized anti-cancer specific

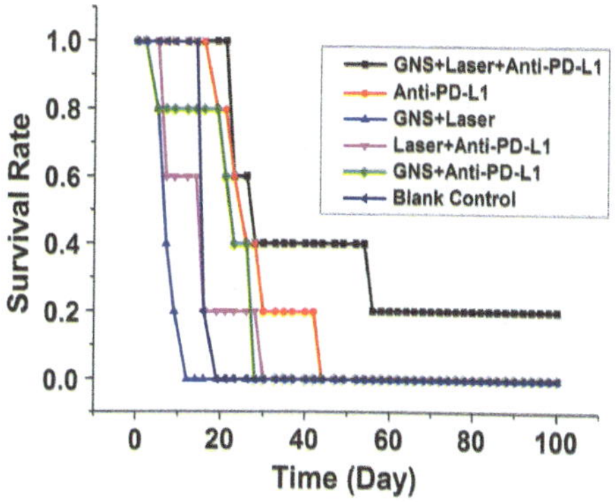

Fig. 6.3 Kaplan-Meier (K-M) overall survival curve. At the end of 49 days, only photoimmunotherapy (GNS + Laser + Anti-PD-L1) group has 2 survival mice (40%) and only this group has one mouse survived after 100 days while all other control groups have none survival mouse. (Adapted from Ref. [25])

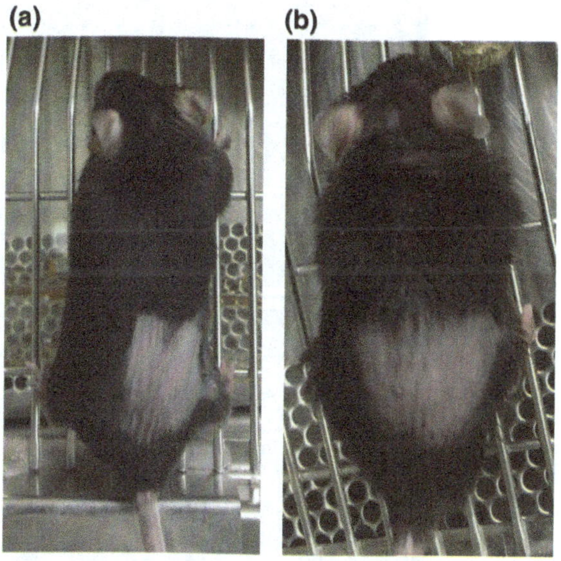

Fig. 6.4 Mice with no tumor developed after cancer rechallenge. After photoimmunotherapy with IV GNS injection (**a**) or intratumoral GNS injection (**b**), both left and right flank tumors disappeared. The two mice were monitored for 3 months and there was no tumor recurrence. The rechallenge was performed by injecting 250,000 MB49 bladder cancer cells under dorsal skin and no tumor developed after two months. (Adapted from Ref. [25])

immunoresponse. The observed long-lasting and specific therapeutic effect from a memorized immune responses is of significant value not only to treat primary tumors and cancer metastasis but also to prevent cancer recurrence.

Furthermore, in order to confirm the triggered immunoresponse following the combined photothermal therapy and immunotherapy, mice with green fluorescent protein (GFP) expression and window chamber were used to observe immune cells accumulation in the distant tumors after GNS-mediated laser ablation on the primary tumor and anti-PD-L1 injection. Two mice with both flank tumor (primary tumor) and tumor under window chamber (remote tumor) were used for this study. For the mouse in the treatment group, 2 mg GNS was IV injected on day -1. The 0.6 W/cm^2 808 nm CW laser treatment was performed on the flank tumor on day 0. The first window chamber imaging was performed prior to laser treatment. 200 μg of anti-PD-L1 antibody were intraperitoneally injected on day 0, day 3 and day 6 after laser treatment. The only difference between the treatment and the control mouse was that the control mouse had no laser treatment on the primary flank tumor. Fluorescence imaging (Fig. 6.5) shows that the distant tumor under window chamber in the treatment mouse has significantly higher fluorescence signal (more immune cells) one week after the photoimmunotherapy. On the contrary, the control mouse with GNS IV injection and anti-PD-L1 IP injection doesn't show increased fluorescence signal with time, indicating no immunoresponse existence.

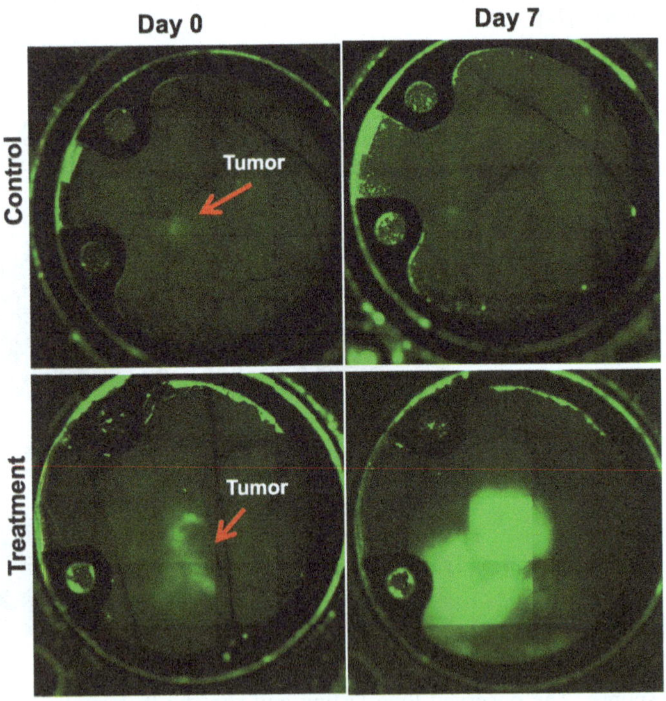

Fig. 6.5 Window chamber imaging for mice with green fluorescence protein expression in immune cells. Tumors were developed in both flank (primary tumor) and under window chamber (distant tumor). For the treatment mouse, 2 mg GNS was IV injected on day -1 and 10 min 0.6 W/cm^2 laser treatment was performed on the flank tumor on day 0. 200 μg anti-PD-L1 antibody was IP injected on day 0, day 3 and day 6 after laser treatment. For the control mouse, GNS and anti-PD-L1 antibody were administrated the same as the treatment mouse but no laser irradiation was performed. It can be seen that green fluorescence signal in the treatment mouse increases significantly one week after laser irradiation on the flank tumor, showing more immune cells accumulation in the distant tumor under window chamber

The observed therapeutic effect of photoimmunotherapy could be explained by the synergistic effect between photothermal therapy and checkpoint inhibitor immunotherapy. Localized PTT with GNS and NIR irradiation selectively and effectively kills primary tumor cells. Given the same mass, GNS nanoparticles have 4–5 orders of magnitude higher photon-to-heat conversion efficiency than water. As a result, with GNS selective accumulation in tumors, PTT can ablate tumor specifically with limited damage to surrounding healthy tissues with water as main NIR light absorber. The ability to perform specific tumor ablation offers significant advantages compared to other thermal therapies. Microwaves, radiofrequency and ultrasound can control macroscopically heating around the tumor region, but cannot precisely ablate cancer cells in a timely manner. After PTT with GNS enhancement, dying tumor cells could release cancer-specific antigens, damage-associated molecular pattern molecules (DAMPs), heat shock proteins (HSPs), etc. DAMPs are intracellular molecules that are normally hidden in live cells. When cells are damaged or dying, DAMPS are released and trigger specific immune response. APCs, which are present in the tissue or in local draining lymph nodes, process the tumor antigens and present tumor-derived peptides to T cells. DAMPS have been shown to exert various effects on antigen-presenting cells (APCs), such as maturation, activation and antigen processing/presentation. The anti-PD-L1 antibodies inhibit a key pathway by which tumor cells escape from immune monitoring. By suppressing this tumor defense mechanism, the tumor cells are now vulnerable to the cytotoxic T cells that have been stimulated by cancer-specific antigens released by GNS-enabled photothermal therapy. Therefore, a combination of photothermal therapy and checkpoint inhibitor immunotherapy could generate synergestic effect with specific cytotoxic T cells to attack vulnerable cancer cells in the body.

6.4 Summary

In summary, photoimmunotherapy, a combination of photothermal therapy and checkpoint inhibitor immunotherapy, shows a synergistic effect of immune triggering resulting from precise tumor ablation with NIR light and tumor immune suppression inversion with checkpoint inhibitor, anti-PD-L1 antibody. Our experimental results show the photoimmunotherapy has great promise to treat not only unresectable primary tumors but also distant cancer metastasis. Furthermore, the identified long-lasting therapeutic effect from memorized immune response also shows great potential to prevent cancer recurrence. Further studies including PTT thermal dose optimization and full immune response mechanism investigation should be explored to optimize our novel photoimmunotherapy for future cancer treatment.

References

1. Ferlay J et al (2015) Cancer incidence and mortality worldwide: sources, methods and major patterns in GLOBOCAN 2012. Int J Cancer 136:E359–E386
2. Pignon JP, le Maitre A, Maillard E, Bourhis J, Grp M-NC (2009) Meta-analysis of chemotherapy in head and neck cancer (MACH-NC): an update on 93 randomised trials and 17,346 patients. Radiother Oncol 92(1):4–14
3. Van Cutsem E et al (2007) Open-label phase III trial of panitumumab plus best supportive care compared with best supportive care alone in patients with chemotherapy-refractory metastatic colorectal cancer. J Clin Oncol 25:1658–1664
4. Gebski V et al (2007) Survival benefits from neoadjuvant chemoradiotherapy or chemotherapy in oesophageal carcinoma: a meta-analysis. Lancet Oncol 8:226–234
5. Wagner AD, Grothe W, Haerting J, Kleber G, Grothey A, Fleig WE (2006) Chemotherapy in advanced gastric cancer: a systematic review and meta-analysis based on aggregate data. J Clin Oncol 24:2903–2909
6. Vauthey JN et al (2006) Chemotherapy regimen predicts steatohepatitis and an increase in 90-day mortality after surgery for hepatic colorectal metastases. J Clin Oncol 24:2065–2072
7. Benson AB et al (2004) American society of clinical oncology recommendations on adjuvant chemotherapy for stage II colon cancer. J Clin Oncol 22:3408–3419
8. Afra D et al (2002) Chemotherapy in adult high-grade glioma: a systematic review and meta-analysis of individual patient data from 12 randomised trials. Lancet 359:1011–1018
9. Takahashi H et al (2013) Prognostic significance of programmed cell death-1-positive cells in follicular lymphoma patients may alter in the rituximab era. Eur J Haematol 90:286–290
10. Lyford-Pike S et al (2013) Evidence for a role of the PD-1:PD-L1 pathway in immune resistance of HPV-associated head and neck squamous cell carcinoma. Can Res 73: 1733–1741
11. Saresella M, Rainone V, Al-Daghri NM, Clerici M, Trabattoni D (2012) The PD-1/PD-L1 pathway in human pathology. Curr Mol Med 12:259–267
12. Zeng Z et al (2011) Upregulation of circulating PD-L1/PD-1 is associated with poor post-cryoablation prognosis in patients with HBV-related hepatocellular carcinoma. Plos One 6
13. Shi F et al (2011) PD-1 and PD-L1 upregulation promotes CD8(+) T-cell apoptosis and postoperative recurrence in hepatocellular carcinoma patients. Int J Cancer 128:887–896
14. Mu C-Y, Huang J-A, Chen Y, Chen C, Zhang X-G (2011) High expression of PD-L1 in lung cancer may contribute to poor prognosis and tumor cells immune escape through suppressing tumor infiltrating dendritic cells maturation. Med Oncol 28:682–688
15. Gadiot J, Hooijkaas AI, Kaiser ADM, van Tinteren H, van Boven H, Blank C (2011) Overall survival and PD-L1 expression in metastasized malignant melanoma. Cancer 117:2192–2201
16. Okazaki T, Honjo T (2007) PD-1 and PD-1 ligands: from discovery to clinical application. Int Immunol 19:813–824
17. Ghebeh H et al (2006) The B7-H1 (PD-L1) T lymphocyte-inhibitory molecule is expressed in breast cancer patients with infiltrating ductal carcinoma: Correlation with important high-risk prognostic factors. Neoplasia 8:190–198
18. Latchman YE et al (2004) PD-L1-deficient mice show that PD-L1 on T cells, antigen-presenting cells, and host tissues negatively regulates T cells. Proc Natl Acad Sci USA 101:10691–10696
19. Curran MA, Montalvo W, Yagita H, Allison JP (2010) PD-1 and CTLA-4 combination blockade expands infiltrating T cells and reduces regulatory T and myeloid cells within B16 melanoma tumors. Proc Natl Acad Sci USA 107:4275–4280
20. Keir ME, Butte MJ, Freeman GJ, Sharpel AH (2008) PD-1 and its ligands in tolerance and immunity. Annu Rev Immunol 26:677–704
21. Hino R et al (2010) Tumor cell expression of programmed cell death-1 ligand 1 is a prognostic factor for malignant melanoma. Cancer 116:1757–1766

22. Zheng Y-W, Li R-M, Zhang X-W, Ren X-B (2013) Current adoptive immunotherapy in non-small cell lung cancer and potential influence of therapy outcome. Cancer Invest 31: 197–205
23. Wolchok JD et al (2013) Nivolumab plus ipilimumab in advanced melanoma. N Engl J Med 369:122–133
24. Hamid O et al (2013) Safety and tumor responses with lambrolizumab (Anti-PD-1) in melanoma. N Engl J Med 369:134–144
25. Liu Y, Maccarini P, Palmer GM, Etienne W, Zhao Y, Lee CT, Ma X, Inman BA, Vo-Dinh T (2017) Synergistic immuno photothermal nanotherapy (SYMPHONY) for the treatment of unresectable and metastatic cancer. Sci Rep 7(8606):1–6

Chapter 7
Conclusion and Future Outlook

7.1 Conclusion

The developed multifunctional GNS theranostic nanoprobe has multiple capabilities ranging from in vivo tracking at the subcellular all the way up to the whole body level as well as specific cancer treatment with photoimmunotherapy, providing a superior nanoplatform for both pre-clinical basic research and clinical translation in cancer management. In vivo tracking and biodistribution studies have shown that dense PEG coating, small particle size and higher injection dose increase GNS accumulation in tumor. With immunohistochemistry analysis, GNS nanoparticles were found to be cleared by macrophage cells in liver and spleen after IV systematical administration. Intraoperative tumor detection was demonstrated as a proof-of-principle with GNS multifunction nanoprobe using the SERS detection method. High-resolution sensitive TPL imaging showed that smaller nanoparticles have deeper tumor penetration depth after leaking through tumor vasculature. ^{124}I labeled GNS nanoprobe with selective accumulation in brain tumor but not in normal brain has been applied for early brain tumor detection down to 0.5 mm in diameter using exceptionally sensitive PET imaging. TEM imaging was also performed to confirm that GNS nanoprobes are endocytosized into brain tumor cells after IV tail vein injection. In addition, the GNS nanoparticles were demonstrated to have much higher photon-to-heat conversion efficiency than water, the main light absorber in tissues. The plasmonic GNS-enhanced photothermal therapy with NIR laser has been shown to produce specific therapeutic effect on tumor with high GNS accumulation. The generated photothermal ablation has been observed to generate immune triggering effect (cancer vaccine), which is combined with checkpoint inhibitor immunotherapy to treat cancer metastasis. It should be noted that, the generated immune response was found to exhibit a memorized long-term effect that

Y. Liu, *Multifunctional Gold Nanostars for Cancer Theranostics*, Springer Theses,
https://doi.org/10.1007/978-3-319-74920-4_7

can prevent tumor development after rechallenge, indicating its potential to be used to prevent cancer recurrence. What is more, the 6-month long-term toxicity study including body weight monitoring, blood chemistry test and histopathology examination demonstrated GNS nanoparticles' biocompatibility. As a result, biocompatible GNS nanoprobes with superior properties for cancer detection and specific therapy have great promise for future clinical translation in cancer management with the ultimate aim to improve patients' overall survival.

7.2 Future Outlook

In this study, the developed multifunctional GNS nanoprobes have demonstrated their great potential for future clinical applications. However, extensive efforts are still required in the future to further investigate several key questions before clinical translation. (1) How to improve GNS uptake in tumor? PEGylated GNS nanoprobes have typically 2–5% ID/g tumor uptake with passive targeting (EPR effect). It might be sufficient for imaging purposes as demonstrated in sensitive brain tumor detection, but it is barely adequate for anti-cancer drug delivery. Active targeting mechanism might be applied to improve tumor accumulation although it is still controversial whether antibodies can improve nanoparticles accumulation in tumors. (2) How to clear GNS nanoprobe from body following injections? Most of GNS nanoprobes in this study were found to be cleared from blood circulation by macrophage cells in liver and spleen after IV injection. Although long-term toxicity has demonstrated GNS nanoprobes are biocompatible, clearable nanoparticles are still preferable for future clinical applications. Nanoparticles with size less than 10 nm can be cleared through the kidney. It would be worth trying to synthesize star-shaped plasmonic nanoparticles with size less than 10 nm. Alternatively, biodegradable GNS could be tried so that they might degrade after application and eventually get cleared from body. (3) How to perform GNS synthesis to meet Good Manufacturing Process (GMP) requirements. It is crucial to synthesize GNS nanoparticles consistently for future commercialization. The automatic synthesizer might be used to control quality and make synthesized GNS nanoparticles consistent across batches. (4) Further in vivo toxicity studies with other animal models. Food and Drug Administration (FDA) requires toxicity results from at least two animal models before clinical trials. Therefore, other animal models such as rats, dogs or pigs could be used to further demonstrate the biocompatibility of GNS. (5) Mechanism investigation of photoimmunotherapy. The observed immune triggering effect like cancer vaccine, is of significant value since it has the potential to provide a novel method to treat cancer metastases as well as to prevent recurrence with memorized specific cytotoxic immune cells. Detailed immune mechanism studies are required to optimize our novel photoimmunotherapy strategy before clinical applications.

The promising results obtained from this study have shown great potential of GNS nanoprobes for future clinical translation. However, we should realize that there is still a long way to go before clinical applications to improve patients' overall survival. It is exactly what we would like to pursue in the coming years.

We choose to fight against cancer, not because it is easy, but hard!

Yang Liu

About the Author

Yang Liu was born in China and obtained his Bachelor (2007) and Master degree (2010) in Chemistry from the Shanghai Jiao Tong University. He studied molecular simulation in Dr. Huai Sun's lab and led a team to win the best performance in the 5th Industrial Fluid Properties Simulation Challenge (IFPSC), organized by the National Institute of Standards and Technology of the United States of America. After coming to the Duke University in 2010, he pursues his graduate studies in Chemistry and Biomedical Engineering, where he developed a multifunctional nanoprobe for sensitive cancer detection and photoimmunotherapy. In the past 6 years, he published 6 first-author and 8 co-author peer-reviewed papers. With hard work, he won several awards including the "Kathleen Zielek" fellowship, "Student Travel Award" in the 2013 SPIE Conference, and "Top Posters" in the 2014 Annual Radiation Oncology and Imaging Program Conference. In addition, he received his Master Degree in Biomedical Engineering in 2014 and Ph.D. Degree in Chemistry in 2016. After graduation, he continues his work in multifunctional nanoprobe aimed for clinical translation as a postdoctoral researcher under co-supervision of Dr. Tuan Vo-Dinh and Dr. Nelson Chao from the Duke University Medical Center starting in January 2017.

Y. Liu, *Multifunctional Gold Nanostars for Cancer Theranostics*, Springer Theses,
https://doi.org/10.1007/978-3-319-74920-4

Provisional Patent

1. Tuan Vo-Dinh, Brant A. Inman, Paolo Maccarini, Greg Palmer, **Yang Liu** and Douglas Weitzel, "Synergistic immuno photo nanotherapy systems and methods of use", US Provisional Patent Application No. 62/182,734, filed on June 22, 2016.

Y. Liu, *Multifunctional Gold Nanostars for Cancer Theranostics*, Springer Theses,
https://doi.org/10.1007/978-3-319-74920-4

The manufacturer's authorised representative in the EU is Springer Nature Customer Service Centre GmbH, Europaplatz 3, 69115 Heidelberg, Germany. If you have any concerns regarding our products, please contact ProductSafety@springernature.com

Printed and bound by CPI Group (UK) Ltd, Croydon, CR0 4YY
05/10/2026
02228152-0004